D1062424

The Institute of Mathematics
and its Applications
Conference Series

The Institute of Mathematics
and its Applications
Conference Series

Previous volumes in this series were published by
Academic Press to whom all enquiries should be addressed.
Forthcoming volumes will be published by
Oxford University Press throughout the world.

NEW SERIES

1. *Supercomputers and parallel computation* Edited by D. J. Paddon
2. *The mathematical basis of finite element methods*
   Edited by David F. Griffiths
3. *Multigrid methods for integral and differential equations*
   Edited by D. J. Paddon and H. Holstein
4. *Turbulence and diffusion in stable environments* Edited by J. C. R. Hunt
5. *Wave propagation and scattering* Edited by B. J. Uscinski
6. *The mathematics of surfaces* Edited by J. A. Gregory

The mathematics of surfaces

The proceedings of a conference
organized by the Institute of Mathematics and its Applications
and held at the University of Manchester, 17–19 September 1984

Edited by

J. A. GREGORY
Brunel University

CLARENDON PRESS · OXFORD · 1986

Oxford University Press, Walton Street, Oxford OX2 6DP

Oxford New York Toronto
Delhi Bombay Calcutta Madras Karachi
Kuala Lumpur Singapore Hong Kong Tokyo
Nairobi Dar es Salaam Cape Town
Melbourne Auckland

and associated companies in
Beirut Berlin Ibadan Nicosia

Oxford is a trademark of Oxford University Press

Published in the United States
by Oxford University Press, New York

British Library Cataloguing in Publication Data
The Mathematics of surfaces: proceedings of a conference organized by the Institute of Mathematics and its Applications and held at the University of Manchester, 17–19 September 1984. – (The Institute of Mathematics and its Applications conference series. New series; 6)
1. Surfaces 2. Geometry
I. Gregory, J.A. II. Institute of Mathematics and its Applications III. Series
516.3'6 QA571
ISBN 0 19 853609 7

Printed in Great Britain by St Edmundsury Press,
Bury St Edmunds, Suffolk

PREFACE

This book contains the proceedings of the conference "The Mathematics of Surfaces" organised by the Institute of Mathematics and its Applications and held at the University of Manchester from 17th-19th September, 1984.

The main aim of the conference was to consider mathematical techniques suitable for the description and analysis of surfaces in three dimensions, and to consider the application of such techniques in areas such as "computer-aided geometric design".

The papers range from those of an introductory nature, to ones of a more advanced or specialist character.

The book begins with expository papers on the basic mathematical tools of computational geometry, classical differential geometry, parametric representations for computer aided design and differential forms. Further papers deal with algorithms for multivariate splines, recursive division techniques, surface-surface intersections, principal surface patches including cyclide surfaces, N-sided patches, Gaussian curvature and shell structures, and flexible surface structures.

The conference organising committee under the Chairmanship of Professor M.J.D. Powell is to be thanked for its help and special thanks are due to Catherine Richards, of the IMA, for overseeing the organisation of the conference and the production of the book. Mrs. S. Hockett and Miss D. Wright, of the IMA, and Dr. T.W. Hewitt, of the University of Manchester, arranged the smooth running of the conference and Miss Pamela Irving, Miss Karen Jenkins, Mrs. Janet Parsons and Mrs. Joanne Robinson typed the manuscripts for the book. Finally, special thanks are due to the contributing authors, for this is their book.

J.A. Gregory
Brunel University

ACKNOWLEDGEMENTS

The Institute thanks the authors of the papers, the editor, Dr. J.A. Gregory (Brunel University) and also Miss Pamela Irving, Miss Karen Jenkins, Mrs. Janet Parsons and Mrs. Joanne Robinson for typing the papers.

CONTENTS

Contributors xi

Introduction to the basic mathematical tools by M.A. Sabin and R.R. Martin 1

Parametric curves and surfaces as used in computer aided design by M.J. Pratt 19

Improvements to parametric bicubic surface patches by D.P. Sturge 47

A circle diagram for local differential geometry by A.W. Nutbourne 59

Elementary exposition of differential forms by L.M. Woodward 73

Differential forms in the theory of surfaces by L.M. Woodward 95

Surface/surface intersection problems by M.J. Pratt and A.D. Geisow 117

Defining and designing curved flexible tensile surface structures by C.J.K. Williams 143

Gaussian curvature and shell structures by C.R. Calladine 179

Multivariate spline algorithms by W. Boehm 197

N-sided surface patches by J.A. Gregory 217

The solution of a frame matching equation by A.W. Nutbourne 233

Cyclide surfaces in computer aided design by R.R. Martin, J. de Pont and T.J. Sharrock 253

Recursive Division by M.A. Sabin 269

CONTRIBUTORS

W. BOEHM; *Angewandte Geometrie und Geometrische EDV, Technische Universitaet Braunschweig, D-3300 Braunschweig, Federal Republic of Germany.*

C.R. CALLADINE; *Cambridge University Engineering Department, Trumpington Street, Cambridge, CB2 1PZ.*

J. DE PONT; *Cambridge University Engineering Department, Trumpington Street, Cambridge, CB2 1PZ.*

A.D. GEISOW; *Cambridge Interactive Systems Limited, 20 Market Street, Swavesey, Cambridge, CB4 5QG.*

J.A. GREGORY; *Department of Mathematics and Statistics, Brunel University, Uxbridge, Middlesex, UB8 3PH.*

R.R. MARTIN; *Department of Computing Mathematics, University College, Senghennydd Road, Cardiff, CF2 4AG, Wales.*

A.W. NUTBOURNE; *Cambridge University Engineering Department, Trumpington Street, Cambridge, CB2 1PZ.*

M.J. PRATT; *Department of Applied Computing and Mathematics, Cranfield Institute of Technology, Cranfield, Bedford, MK43 0AL.*

M.A. SABIN; *Fegs Limited, Oakington, Cambridge, CB4 5BA.*

T.J. SHARROCK; *Cambridge University Engineering Department, Trumpington Street, Cambridge, CB2 1PZ.*

D.P. STURGE; *Delta C.A.E. Limited, 20 Trumpington Street, Cambridge, CB2 1QA.*

C.J.K. WILLIAMS; *School of Architecture & Building Engineering, University of Bath, Claverton Down, Bath, BA2 7AY.*

L.M. WOODWARD; *Department of Mathematical Sciences, University of Durham, South Road, Durham, DH1 3LE.*

INTRODUCTION TO THE BASIC MATHEMATICAL TOOLS

M.A. Sabin
(Fegs Ltd., Cambridge)

R.R. Martin
(University College, Cardiff)

The purpose of this introduction is to bridge the gap between the mainstream of mathematical methods likely to be familiar to all those reading the book and the techniques used as an everyday basis by those working in the world of numerical representations of smooth surfaces.

The items to be discussed are:-

1. Context and Objectives
2. Definitions and Representations: Interrogation
3. Coordinate systems
4. Vector notation, Tensors and Quarternions
5. Explicit, Implicit and Parametric representations
6. Differential Geometry
7. Duality, Plane coordinates, Envelopes and Pedal equations
8. Basis functions, Tensor products
9. Piecewise forms

1. CONTEXT AND OBJECTIVES

The objective of numerical geometry is to be able to define specific shapes of things - as specific as the shape of a particular make of car, or a specific size and fitting of shoe - and then to use numerical procedures to determine properties of those shapes.

In this book we shall discuss various methods of describing shapes. Each method can be instantiated into many specific shapes by the selection of numerical coeffcicents. Those coefficients which can be directly controlled are informally referred to as 'handles', which are 'turned' until the shape is the desired one.

Once a method of shape description has been chosen, algorithms can be written once and for all, so that new instances can have their properties determined without writing new software.

Operational Criteria

Criteria for the choice of a method include:

(i) The ease of choosing values for the handles to achieve a desired shape - either initially or by perturbation from a previous definition.

(ii) The robustness of algorithms using that method.

(iii) The computing economy of the method in terms of the amount of information needing to be stored, or in terms of the speed of the algorithms accessing it.

Elegance of formulation is not itself an operational criterion, but more often than not an elegant form gives a method which competes well.

2. DEFINITIONS AND REPRESENTATIONS: INTERROGATION

This section deals with two concepts which are easily confused. We distinguish between the definition and representation of a surface as follows:

The DEFINITION of a specific surface is the information which is provided by the user when an instance is selected. The surface designer has certain options and possibilities consciously in mind.

The REPRESENTATION of a surface is the set of numbers held within the computing machine which distinguish that instance from all other possible instances.

For a particular method the universe of shapes given by all possible representations must obviously include that of all possible definitions, but it can be much broader (occupying more store) in order to permit faster algorithms.

Considerable amounts of processing can go on between a definition and the corresponding representation, in order to get the best combination of criteria (i), (ii) and (iii) above.

INTERROGATION is a generic term for all algorithms which compute properties of a surface. Common examples are

(i) drawing some image of the surface

(ii) plane cross sections

(iii) intersections of two surfaces

(iv) silhouette curves

(v) the area of a region of a surface and

(vi) the cutter path for machining a die which will form the surface

However this list is not exclusive or complete.

An important point to note is that the available interrogation facilities can make a great deal of difference in the ease of definition of a surface. It is not easy to understand the shape of a surface displayed only by its lines of constant parameter (see Section 5 below), whereas a nest of closely spaced plane sections can be interpreted by the eye as a contour map. Loftsmen are particularly skilled in this kind of interpretation. B-spline methods (see the next paper) have a control network which exaggerates certain features of the shape thus making those features easier to detect.

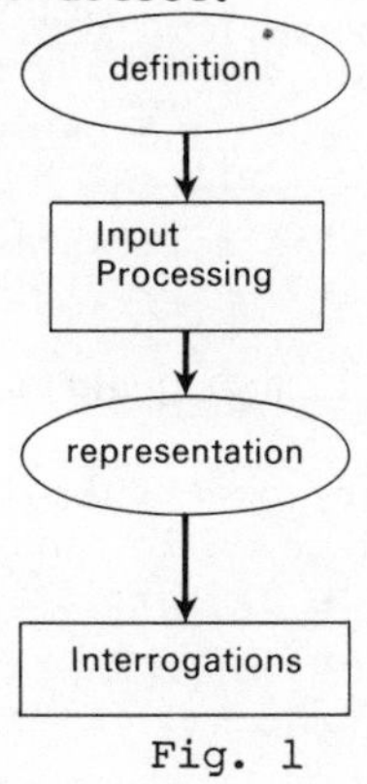

Fig. 1

3. COORDINATE SYSTEMS

Cartesian coordinates are part of most school syllabi, and so this topic needs little revision.

A three dimensional rectangular cartesian coordinate system consists of three perpendicular directed lines, called the <u>axes</u>, which intersect at a reference position called the <u>origin</u>. The displacement from the origin to any point in space can be represented as the sum of three displacements along directions parallel to each of the axes. The signed magnitudes of these displacements are called the cartesian coordinates. Let unit displacements along each axis be represented by (1, 0, 0), (0, 1, 0) and (0, 0, 1) respectively. Then a general point with displacements x, y and z may be represented by the composition

$$(x,y,z) = (0,0,0) + x(1,0,0) + y(0,1,0) + z(0,0,1)$$

where (0,0,0) denotes the origin.

A local coordinate system, being composed of a position and three displacements can itself be represented with respect to the "world" or implicit coordinate system. The algorithms are important which, given the coordinates of a position with respect to a local coordinate system, compute the coordinates in the implicit system, and vice versa.

Non-Cartesian coordinates

School syllabi also cover coordinate systems other than cartesian: polar, spherical polar, etc. These are used in mainstream surface work for two purposes only - if their use enables more comfortable definition, or if their use gives a reduction in dimensionality of the entire problem to be solved. The most general curvilinear coordinate system, not usually explored at school in any depth, is however a tool of regular use. The parametric surface definition of Section 5 is just an embedding of a general 2D curvilinear coordinate system (parameter space) in a cartesian 3D system.

4. VECTOR NOTATION, TENSORS AND QUARTERNIONS

Vector notation is a very powerful tool for keeping down the amount of algebraic manipulation in surface work. It is simply the device of using a single algebraic symbol for a position or a displacement.

Dot and Cross products

Part of this economy is in the writing of one equation instead of two or three. More savings are encountered as a result of the recognition of two common constructions and labelling them the vector DOT product (or inner product) and the vector CROSS product.

If

$$\underline{v1} = (x1, y1, z1) \text{ and } \underline{v2} = (x2, y2, z2)$$

then their dot product is a scalar

$$\underline{v1} \cdot \underline{v2} = x1*x2+y1*y2+z1*z2$$

Their cross product is another vector

$$\underline{v1} \times \underline{v2} = (y1*z2-y2*z1,\ z1*x2-z2*x1,\ x1*y2-x2*y1)$$

The dot product is largest when $\underline{v1}$ and $\underline{v2}$ are parallel and zero when they are perpendicular. The cross product is zero when they are parallel and maximum when they are perpendicular. The cross product has a direction perpendicular to $\underline{v1}$ and $\underline{v2}$.

Vector Differentiation

Once a position is denoted by a single symbol and we start writing algebra, it becomes tempting to extend the notation to include differentiation.

Differentiation of a vector $\underline{v}$ with respect to some variable is just the vector whose components are the derivatives of the components of $\underline{v}$.

If $$\underline{v} = (x,y,z)$$

then $d\underline{v}/dq = (dx/dq,\ dy/dq,\ dz/dq)$.

Differentiation with respect to a vector is slightly less intuitive. However, to a first order of approximation we have

$$f(\underline{p} + d\underline{p}) = f(\underline{p}) + dx\,\frac{\partial f}{\partial x} + dy\,\frac{\partial f}{\partial y} + dz\,\frac{\partial f}{\partial z}\,.$$

This can be written as

$$f(\underline{p} + d\underline{p}) = f(\underline{p}) + d\underline{p}.\,\frac{df}{d\underline{p}}\,,$$

where we define

$$\frac{df}{dp} = \left(\frac{\partial f}{\partial x}\,,\ \frac{\partial f}{\partial y}\,,\ \frac{\partial f}{\partial z}\right).$$

This latter quantity is called the gradient of f and is commonly denoted by ∇f.

Tensors

Tensor notation may be regarded as an extension of vectors. This also uses a single symbol for a geometric entity, but indicates rather more explicitly that it has components.

The symbol P^i indicates that P has components P^1, P^2 and P^3 and that these components are measured in units of length. The symbol P_i indicates that the components are P_1, P_2 and P_3, measured in spatial frequencies (per-inch or per-metre).

The exact equivalent of the dot product is the inner product of two tensors

$$P^i Q_i$$

where the convention is assumed that if the same letter is used as a superscript and a subscript in the same term of an expression, then it will be given all values in turn and the resulting products summed (the summation convention).

Latin super- and sub-scripts (i, j, k ...) are usually taken to imply variation over the three space dimensions, greek letters (α, β, γ ...) to imply variation over the two curvilinear coordinates of a surface.

This notation has the advantage of generalising into more complex objects. A rotation matrix, for example, can be represented by the symbol

$$A^i_j$$

which can have a dot product on both sides with two vectors

$$P_i A^i_j Q^j$$

The cross product can be represented as a product involving a special tensor, the alternating tensor. For example, the surface normal may be given by the product of two tangent vectors

$$N_k = e_{ijk} T_1^i T_2^j$$

The use of subscripts for the surface normal emphasises its relationship with the tangent plane, whose equation in tensor notation is

$$(P^i - Q^i) N_i = 0$$

Quarternions

Quarternions are a generalisation of complex numbers; they were invented in the last century by Hamilton, and popular for some time, before vector methods finally gained supremacy over them. Nevertheless, quarternions are better than vectors for solving certain problems in geometry, especially where rotations are concerned, and so interest in them has recently revived.

Whereas a complex number may be written in the form

$$z = a + ib \quad \text{where } i^2 = -1$$

a quarternion may be written as

$$q = a + ib + jc + kd \quad \text{where } i^2 = j^2 = k^2 = -1$$

and furthermore

$$ij = k,\ jk = i,\ ki = j$$
$$\text{and } ji = -k,\ kj = -i,\ ik = -j$$

Quarternions are added and subtracted in the obvious way, component by component, while the above relationships are used when performing multiplication . However, because of the asymmetry of the above, the product qr is not in general the same as the product rq. Thus just as for matrices, for which there is a similar asymmetry, it is not possible to define quarternion division. Instead we can define an inverse quarternion q^{-1} for each quarternion q such that $q\,q^{-1} = q^{-1}q = 1$.

To convert a problem from vector notation to quarternions we replace each vector ($\underline{v}$ = (a, b, c) say) by the quarternion q = ia + jb + kc. (Note that the 'real' component is zero.) If two such quarternions, q and r say, representing vectors are multiplied together, the result has a real part and a 'vector' part. The result can loosely be written

$$qr = -\,\underline{q}.\underline{r} + \underline{q} \times \underline{r}$$

Thus the quaternion product includes both the scalar and the vector product as parts of its result.

If we wish to rotate a given vector represented by the quarternion q through an angle θ about an axis with direction cosines (l, m, n) we define a quarternion r such that

$$\begin{aligned} r = {} & \cos(\theta/2) \\ & + i\,l\,\sin(\theta/2) \\ & + j\,m\,\sin(\theta/2) \\ & + k\,m\,\sin(\theta/2) \end{aligned}$$

and find the new orientation q' of q by evaluating

$$q' = r\,qr^{-1}$$

5. EXPLICIT, IMPLICIT AND PARAMETRIC REPRESENTATIONS

The most familiar way of associating mathematics with shape, again from school days, is by the use of graphs. The graph of some relationship between the coordinates of a point gives a graphic representation of a shape. At school one probably looked only at the explicit relationships, of the form

$$y = f(x) \text{ or possibly } z = f(x,y)$$

It is certainly possible to represent a shape of a limited class by holding the coefficients of, say, a polynomial. This hits problems, though, in that one property of real shapes is that they do not alter when you move them around and rotate them. Still less do they vanish. Unfortunately, explicit curves and surfaces do not remain explicit when you rotate them. Either you have to approximate or else you lose the explicit form. Further, representation of shapes with 'vertical' slopes needs special treatment which has to be rather ad-hoc.

Two alternatives which do not suffer from these problems are the Implicit form

$$f(x,y,z) = 0$$

and the Parametric form

$$\underline{p} = \underline{f}(u,v)$$

The implicit form turns out not to have such convenient handles and so the parametric representation is central to all mainstream surface work of the past two decades.

An example of the advantage of the parametric form is that a surface which is defined as the offset from a parametric surface is itself parametric, and while the expression for its parametric function may not appear particularly simple algebraically, it can be coded straightforwardly as an algorithm for calculating the position and its derivatives from those of the basic surface.

6. DIFFERENTIAL GEOMETRY

The aim of this section is to give an intuitive feel to some of the main concepts of differential geometry, with a minimum of equations.

2D Curves

Consider first a curve in two dimensions. This may be represented explicitly as $y=f(x)$, implicitly as $f(x,y)=0$ or parametrically as $\underline{p}=\underline{f}(u)$. It may also be represented in terms of the way the curvature varies along the curve. For this purpose the intrinsic parametrisation is used, in which arc length, usually denoted by s, is used as the free variable. The arc length is just the length measured along the curve from some fixed starting point.

With this parametrisation the first derivative $d\underline{p}/ds$ is a unit vector whose direction is that of the tangent to the curve. This unit vector is denoted by $\underline{t}$.

At any point on a smooth curve, a sufficiently small piece of the curve will look like a piece of circle. If the radius of this circle is ρ then the curvature of the curve at this point, denoted by κ is given by $\kappa = 1/\rho$.

The curve normal, denoted by $\underline{n}$ is defined as the unit vector from the given point towards the centre of the circle.

The intrinsic equation for the curve is of the form

$$\kappa = f(s)$$

For example, κ = constant gives a circle, while $\kappa = a * s + b$ gives a Cornu Spiral, used in French Curves.

Another way of looking at curvature is to consider the angle ψ between the tangent to the curve and a fixed line. The curvature is the rate of change of this angle with respect to arc length, thus

$$\kappa = d(\psi)/ds$$

3D Curves

These ideas may be extended from two dimensions to curves in three dimensions, again by considering a very small piece of curve and the circle which most closely approximates it. The plane in which this circle lies is called the osculating plane.

This allows us to define a normal and a curvature for a 3D curve. However, unless the whole of the curve lies in a single plane, the osculating plane varies as the chosen point moves along the curve. The rotation at any point must be about the tangent at that point, and rate of angular rotation with respect to arc length is called the torsion τ of the curve. A plane curve has zero torsion, while a curve with both curvature and torsion constant is a helix.

In three dimensions, just as in two, the normal $\underline{n}$ is at right angles to the tangent $\underline{t}$. We can take a third unit vector at right angles to both of them which is called the curve binormal $\underline{b} = \underline{t} \times \underline{n}$.

These three orthogonal unit vectors make up a 'frame' the orientation of which will change as we go along the curve. The rates of change of its components are given by the Serret-Frenet equations which relate the derivatives of $\underline{t}$, $\underline{n}$ and $\underline{b}$ with respect to s, to the curvature and torsion of the curve

$$d\underline{t}/ds = \kappa\,\underline{n}$$
$$d\underline{n}/ds = \tau\,\underline{b} - \kappa\,\underline{t}$$
$$d\underline{b}/ds = -\,\tau\,\underline{n}$$

Surfaces

The parametric form of a surface equation is

$$\underline{p} = \underline{f}(u,v)$$

If the value of u is fixed, and we vary v, the point $\underline{p}$ traces out a curve in the surface. The derivative $\partial\underline{p}/\partial v$ is a vector tangent to that curve. Similarly, fixing v and varying u gives another curve, whose tangent is $\partial\underline{p}/\partial u$. These two partial derivatives at any point normally give two distinct tangent directions, both of which lie in the tangent plane to the surface at that point. The normal to this plane, which is also normal to the surface at this point, may be evaluated by taking the cross product of the two tangent vectors

$$\underline{N} = \partial\underline{p}/du \times \partial\underline{p}/dv\,.$$

If the point lies on two surfaces with distinct surface normals, the tangent to the curve of intersection of the two surfaces is given by the cross product of the two normal vectors

$$\underline{t} = \underline{N1} \times \underline{N2}\ .$$

A curve frame was defined above. It is also possible to define a surface frame consisting of two perpendicular tangent vectors and the surface normal. The partial derivatives are not in general perpendicular, but it is always possible to evaluate a second tangent, perpendicular to the first and to the surface normal by yet another cross product.

Surface frames are particularly useful when we are considering the relationship between a surface and a curve lying in it. The first tangent of such a frame would then typically be the tangent to the curve.

Surface Curvatures

The curvature of a surface is rather more complicated than the curvature of a curve, since the curvature of a path across a surface depends on that path as well as on the surface itself.

By taking only paths which are locally normal plane sections of the surface, ie. a plane containing the surface normal, we can reduce the possibilities. The curvature of such a path is the normal curvature in the direction of the tangent to that path.

Different paths passing through the same point in different directions will have different normal curvatures, but it is found that the normal curvature takes on maximum and minimum values for two directions which are always at right angles to each other. These two directions are called principal directions and their values of normal curvature are the principal curvatures, normally denoted by k_1 and k_2.

Any path always tangent to a principal direction is called a line of curvature.

There are two useful scalar measures of curvature, neither of which tells the whole story. The first is the mean curvature denoted by J, the second the Gaussian curvature K

$$J = (k_1 + k_2)/2, \quad K = k_1 k_2$$

Note that if the surface is concave or convex K is positive, while if the surface is saddle shaped K is negative.

First and Second Fundamental Forms

The intrinsic parametrisation of curves has no simple equivalent for surfaces. We have to deal with the more general parametrisation for labelling the various points of the surface. In calculating properties of the surface, such as the angle between the two isoparametric curves through a point, or the normal curvature in a given direction, the coefficients of the first and second fundamental forms are useful.

The first fundamental form is concerned with the first derivatives of the surface. Its coefficients, usually denoted by E, F and G, are obtained by taking the dot products of the two surface tangents. Using $\underline{p}_u$ and $\underline{p}_v$ to denote $\partial\underline{p}/\partial u$ and $\partial\underline{p}/\partial v$

$$E = \underline{p}_u \cdot \underline{p}_u$$

$$F = \underline{p}_u \cdot \underline{p}_v$$

$$G = \underline{p}_v \cdot \underline{p}_v$$

E tells us how far we travel across the surface by varying u, G by varying v. The ratio of F to E and G tells what the angle between the isoparametric curves is.

The second fundamental form is concerned with the second derivatives. Its coefficients, called L, M and N are obtained by taking the dot products of the second derivatives with the surface normal. Writing, for example $\underline{p}_{uu}$ for $\partial^2 \underline{p}/\partial u^2$

$$L = \underline{p}_{uu} \cdot \underline{n}$$

$$M = \underline{p}_{uv} \cdot n$$

$$N = \underline{p}_{vv} \cdot \underline{n}$$

The meanings of these coefficients are not so intuitive as those of the first fundamental form, but they are closely related to the surface normal curvature.

Just as a curve can be specified in shape by defining how its curvature and torsion vary with arc length, a surface can be specified by defining how the coefficients of its first and second fundamental forms vary with u and v. However, these cannot all be chosen completely independently. They must satisfy some consistency equations, known as Gauss's Equation and the Mainardi-Codazzi Equations. These equations relate the coefficients to the Christoffel Symbols which are just the dot products between the first and second derivatives of $\underline{p}$ with respect to u and v.

Curves on Surfaces

If we take a general curve lying in a surface, the curve normal and the surface normal will be distinct. The curvature of the curve can be split into two components. The first component can be regarded as due to the surface curvature. Take the normal plane section which has the same tangent as the curve, and project the curve on to it. Clearly the projection has the surface normal curvature in that direction.

Projection of the curve on to the tangent plane gives the other component, which can be regarded as due to the way the curve bends within the surface. The curvature of this projection is termed the geodesic curvature and is usually denoted by g.

The two components of curvature are related by

$$\kappa^2 = n^2 + g^2 .$$

A curve whose geodesic curvature is everywhere zero is called a geodesic.

7. DUALITY, PLANE COORDINATES, ENVELOPES AND PEDAL EQUATIONS

Duality

One of the more beautiful parts of Euclidean Geometry (the sort of geometry where you prove things without using coordinates) deals with the properties of duality. If there is a theorem which deals with points in a plane and the lines joining them, there is another theorem dealing with lines in the plane and the points in which they intersect. The two theorems are quite distinct, but form an exact pair.

There is a similar symmetry between points and planes in three dimensions, with theorems appearing in pairs. Their proofs match exactly too, since they may be based on the symmetry of the axioms of incidence between points and planes. This may have been one of the stimuli for the modern axiomatic approach.

Plane coordinates

One can go further than this, however. Not only are the axioms symmetric: so too can be the numerical representations, and if this choice is made, identical numerical constructions implement two dual geometric constructions.

The easiest example is the symmetry of points and planes in homogeneous coordinates, where points are represented by quadruples of numbers only the ratios of which are important. These are often normalised by multiplying the quadruple by some factor so that the fourth component is unity. The first three components then correspond to the three space coordinates.

The equation of a plane also has four coefficients, often expressed in the form

$$a*x + b*y + c*z + d = 0$$

If the left hand side is regarded as the inner product of (a,b,c,d) with (x,y,z,1) the symmetry between points and planes is very apparent. Such planes are usually normalised so that the sum of the squares of a,b and c is unity. The fourth component, d, then gives the distance from the origin to the represented plane.

Envelope Surfaces

Once this view is taken, it is a small step to enquire what happens if we generalise the above equation viewed as the condition on a variable plane that it should pass through a fixed point. It turns out that we can set up plane equations of both the implicit and parametric kind, describing bivariate systems of planes. These systems envelope a surface, and thus provide a further method of surface definition.

Since the tangent plane of such a surface is the dual of a point on a conventional surface, the procedure for determining points on an envelope is the dual of (i.e. has the same calculations as) the construction of the tangent plane by differentiation.

The dual of a plane cross section (those points which lie in a given plane) is the silhouette cone (those planes which pass through a given point).

If a surface is defined as a locus, the tangent plane usually appears as a relatively complicated expression. If a surface is defined as an envelope the tangent planes are simpler than the points. By some magic of algebraic geometry it turns out that if a surface has the same (low) order of equation in both forms, it becomes simple in many other aspects too. The nice properties of the cyclide (see paper by Martin, de Pont and Sharrock) can be viewed as stemming from the fact that the dual of a cyclide is another cyclide.

Pedal Equation

A particular parametrisation which leads to interesting properties treats the first three components of the plane representation as the parameters, using the normalisation to reduce the three independent values to two. This implements the mapping of the enveloped surface on to the unit sphere. The entire surface representation then becomes a scalar function (the fourth component) over the unit sphere.

$$h = f(\underline{N})$$

For example, the equation of the sphere with centre $\underline{c}$ and radius r is just

$$h = r + \underline{N} \cdot \underline{c}$$

and that of a convex polyhedron whose vertices are $\underline{p}_i$, i = 1,n

$$h = \max_{i=1,n} (\underline{p}_i \cdot \underline{N})$$

If two surfaces given by such equations slide over each other without rotation, the locus of a point fixed with respect to the moving surface lies in the surface whose equation is

$$h(\underline{N}) = h_1(\underline{N}) + h_2(-\underline{N})$$

This can be turned to good account in problems of interference between moving objects.

8. BASIS FUNCTIONS

In Section 1 above, ease of interactive definition was stressed, whereby the 'handles' are adjusted until the surface being designed has all the desired properties. When comparing surface classes from this point of view, a natural question is "what effect does a small movement of one handle have?"

Many of the surfaces to be considered in this book have the linear form

$$\underline{P}(u,v) = \sum_i \underline{A}_i f_i(u,v)$$

with scalar functions $f_i(u,v)$ and vector coefficients $\underline{A}_i$. In this case, the scalar functions do indeed describe the varying effect over the surface of a unit displacement of the corresponding vector handle. If the f_i behave smoothly, so will the surface: if they are of finite support, control will be local: if they are all non-negative and sum to unity, the surface will lie in the convex hull of its control points.

These functions form a basis for the vector space of surfaces, and so we refer to them as the basis functions of a class of surfaces. It is of course quite possible to have several different bases for the same vector space, in which case it will be possible to convert a surface, without change of shape, from one representation to another, by calculating new coefficients from linear combinations of the old.

Tensor Products

The form

$$\underline{P}(u,v) = \sum_i \underline{A}_i f_i(u,v)$$

is rather too general for practical purposes. The special case normally encountered is the Tensor Product, in which the bivariate basis functions are formed by taking every possible pair one from one set of univariate functions, the other from another. Thus

$$\underline{P}(u,v) = \sum_j \sum_i \underline{A}_{ij}\ f_i(u)\ g_j(v)$$

This form allows us to take a bipolynomial with different degrees in the two directions; for example, to make a ruled surface with curved directrices. Often, however, the set of functions f will be the same set as that of functions g, giving a surface with considerable symmetry between the two directions.

9. PIECEWISE FORMS

The final concept to be described in this introduction is that of piecewise forms.

An important operational property of a surface scheme is that of local modifiability. Having designed the shape of a product, it may be necessary to alter part of that shape to meet changed requirements. Also it may be easier to design the short wavelength features of a surface independently in various parts of the shape.

There are two ways of approaching this:-

* by designing separate "patches" taking due regard to all the continuity requirements between the patches.

* by using piecewise basis functions to generate a whole 'carpet' in one unit.

The two may be exemplified by the problem of designing a polygon. In one case the individual edges are each designed, keeping in mind the need for the end of one edge to have a common position with the start of the next. In the other case the piecewise basis functions shown in Figure 2 are used with coefficients the vertices of the polygon. Because the basis functions have continuity of position the curve being designed does too.

The same pair of alternatives applies to higher order curves and surfaces.

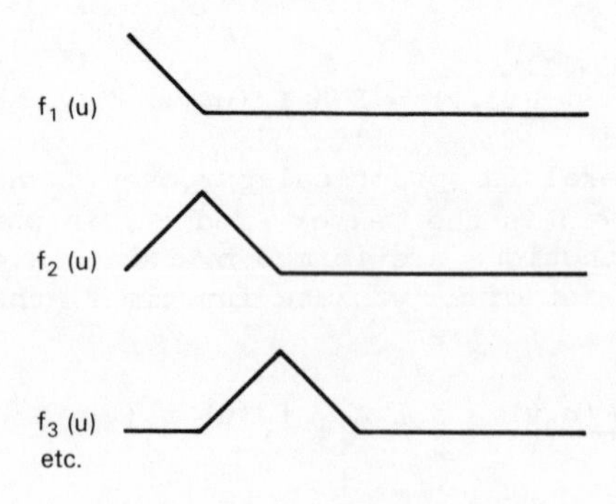

Fig. 2

SOME RECOMMENDED TEXT BOOKS

Vectors:

Rutherford, D.E., *Vectors Methods,* Oliver and Boyd, 1957.

Armit, A.P., *Advanced Level Vectors,* Heinemann, 1968.

Macbeath, A.M., *Elementary Vector Algebra,* OUP, 1966.

Tensors:

Spain, B., *Tensor Calculus,* Oliver and Boyd, 1960.

Jeffreys, H., *Cartesian Tensors,* OUP, 1969.

Quarternions:

Brand, L., *Vector and Tensor Analysis,* Wiley, 1947.

Differential Geometry:

Weatherburn, C.E., *Differential Geometry of Three Dimensions,* CUP, 1927.

Struik, D.J., *Differential Geometry,* Addison Wesley, 1950.

Lipschitz, M.M., *Differential Geometry,* Schaum's Outline Series, McGraw Hill, 1969.

Plane Coordinates:

Maxwell, E.A., *General Homogeneous Coordinates in Space of Three Dimensions,* CUP, 1961.

Envelopes:

Lockwood, E.H., *A Book of Curves,* CUP, 1971.

Algebraic Geometry:

Primrose, E.J.F., *Plane Algebraic Curves,* MacMillan, 1955.

Geometry:

Coolidge, J.L., *A History of Geometrical Methods,* Dover, 1963.

PARAMETRIC CURVES AND SURFACES AS USED IN COMPUTER AIDED DESIGN

M.J. Pratt
(Cranfield Institute of Technology)

1. INTRODUCTION

This paper is intended to provide an introduction to parametric geometry as it is conventionally used in the computer aided design of free-form curves and surfaces. The primary reason for the choice of parametric represention in this context is that it is possible to express curves and surfaces in terms of linear combinations of scalar functions of the parameters, with vector-valued coefficients. Any transformation of the geometry may then be achieved simply by applying the transformation appropriately to the vector coefficients. The mathematical formulation of the geometry remains otherwise unchanged; this is appropriate since, as pointed out by Forrest [16], 'Shape is independent of frame of reference'.

There are other advantages of the parametric formulation, many of which will be illustrated in the remainder of the paper. They include the following:

(i) representations are explicit - it is therefore easy to compute points on curves and surfaces;

(ii) there are no problems in dealing with vertical tangents, for example in the case of smooth closed curves;

(iii) such formulations are conveniently adapted for the definition of curves and surfaces in a piecewise manner;

(iv) it is comparatively simple to compute points on parallel offset curves and surfaces, which is useful in many practical situations.

Finally, it should be emphasised that the curve and surface representations to be described are particularly suited to the construction or synthesis of geometry to meet constraints imposed by the user or designer. These constraints may be either quantitative or aesthetic.

They are presented to the user in terms of what have been called 'handles', i.e. those entities which the user has at his disposal for manipulation leading to the desired geometric result. The 'handles' in any particular geometric representation are precisely the set of vector-valued coefficients referred to earlier. Since a computer aided design system is intended for use by designers rather than mathematicians it is important that these coefficients have some significance which is intuitively clear to the user. This has been achieved in various ways, as will be shown in what follows.

For various reasons, not least their computational convenience, polynomial functions of the parameters are widely used. In the early part of the paper attention will be restricted to the case of cubic curves and bicubic surfaces, though generalisations using polynomials of higher degree are mentioned later. Most of the topics dealt with are treated in greater detail in Faux and Pratt [14]. Original references to individual curve and surface formulations are provided throughout the paper.

2. PARAMETRIC CUBIC CURVES

2.1 The Cubic Hermite (Ferguson) Curve Segment

A parametric cubic curve may be expressed as

$$\underline{r}(t) = \underline{a}_o + \underline{a}_1 t + \underline{a}_2 t^2 + \underline{a}_3 t^3 \tag{2.1}$$

in terms of a parameter t and four vector-valued coefficients $\underline{a}_o$, $\underline{a}_1$, $\underline{a}_2$, $\underline{a}_3$. For practical purposes it is usual to consider only the segment defined by $0 \leq t \leq 1$ and to ignore the behaviour outside this range.

At the end points of the curve segment thus defined we have

$$\begin{aligned} \underline{r}(0) &= \underline{a}_o \\ \underline{r}(1) &= \underline{a}_o + \underline{a}_1 + \underline{a}_2 + \underline{a}_3 \\ \dot{\underline{r}}(0) &= \underline{a}_1 \\ \dot{\underline{r}}(1) &= \underline{a}_1 + 2\underline{a}_2 + 3\underline{a}_3, \end{aligned} \tag{2.2}$$

where a dot denotes differentiation with respect to the parameter t. It is clear that of the 'handles' in equation (2.1) only $\underline{a}_o$ and $\underline{a}_1$ have an obvious geometrical significance.

It would not be reasonable to expect a designer to modify a curve by manipulating $\underline{a}_2$ and $\underline{a}_3$ since the effects of doing this are not intuitively predictable. In the case of $\underline{a}_o$ and $\underline{a}_1$, however, the effects would be those of modifying one end point of the curve segment and its associated tangent vector. These are operations which have a straightforward geometrical interpretation.

Accordingly, it is tempting to reformulate the curve segment so that it can be defined in terms of end points and tangent vectors at both extremities. Set $\underline{r}(0) = \underline{p}(0)$, $\underline{r}(1) = \underline{p}(1)$, $\dot{\underline{r}}(0) = \dot{\underline{p}}(0)$ and $\dot{\underline{r}}(1) = \dot{\underline{p}}(1)$, where $\underline{p}$, $\dot{\underline{p}}$ denote point and vector quantities specified by the user. With these replacements, equation (2.2) may be solved for the $\underline{a}_i$ and the results substituted in (2.1). Rearrangement then yields

$$\underline{r}(t) = \underline{p}(0)\{1 - 3t^2 + 2t^3\} + \underline{p}(1)\{3t^2 - 2t^3\} + \dot{\underline{p}}(0)\{t - 2t^2 + t^3\} + \dot{\underline{p}}(1)\{-t^2 + t^3\}, \quad 0 \leqslant t \leqslant 1. \tag{2.3}$$

In this form the 'handles' have a clear significance for the user. The cubic functions of t are in fact the basis functions for cubic Hermite interpolation. This is hardly surprising, since the construction of a curve segment in terms of end points and tangents in three dimensions is merely a vector analogue of that type of interpolation. It was apparently first described by Ferguson [15]. This is the first example of how a judicious choice of basis functions for the space P_3 of cubic polynomials can provide the user with convenient means of manipulating curves for design purposes.

It is often convenient to express curve segments of this and related types in matrix form. Equation (2.3) may be written as

$$\underline{r}(t) = [1\ t\ t^2\ t^3]\begin{bmatrix} 1 & 0 & 0 & 0 \\ 0 & 0 & 1 & 0 \\ -3 & 3 & -2 & -1 \\ 2 & -2 & 1 & 1 \end{bmatrix}\begin{bmatrix} \underline{p}(0) \\ \underline{p}(1) \\ \dot{\underline{p}}(0) \\ \dot{\underline{p}}(1) \end{bmatrix}, \quad 0 \leqslant t \leqslant 1. \tag{2.4}$$

Here the constant matrix generates the Hermite basis from the standard basis of powers of t, while the vector coefficients or 'handles' occur in the column vector on the right.

One problem with this type of curve formulation is that, while the geometrical significance of the directions of $\dot{\underline{p}}(0)$ and $\dot{\underline{p}}(1)$ is obvious, the interpretation to be placed on their magnitudes is less clear. Experience shows that these magnitudes control what is known as the 'fullness' of the curve. If they are both increased simultaneously the arc length of the curve between the fixed end points increases, and the curve segment bows away from the chord line. Indeed, if the magnitudes are increased sufficiently a loop will occur in a planar curve, as shown in Fig. 2.1.

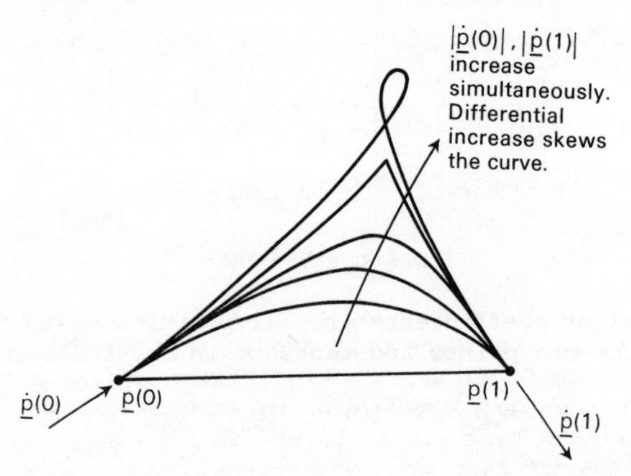

Fig. 2.1: Effect of Tangent Vector Magnitudes on a Ferguson Curve

If the tangent vector magnitudes are changed differentially the curve will be skewed away from the end point where the increase in magnitude is greatest.

The phenomena described in the last paragraph are peculiar to the vector-valued form of Hermite interpolation. They arise because the tangent direction depends only on the ratios of $\dot{x}(t)$, $\dot{y}(t)$ and $\dot{z}(t)$, and so can be the same for infinitely many values of these derivatives provided the ratios are preserved. The magnitude of $\dot{r}(t)$, however, is $\sqrt{(\dot{x}^2(t) + \dot{y}^2(t) + \dot{z}^2(t))}$, and therefore increases or decreases as the three component derivatives increase or decrease together. Thus the curve segment has two degrees of freedom once its end points and end tangent directions are fixed. The tangent vector magnitudes may be made available to the user of a design system as curve adjustment parameters, but not all designers are comfortable with this mode of manipulation. Consequently, other means of defining parametric cubic curves have been devised, notably the Bezier and B-spline formulations which are the preferred choices in many modern design systems.

2.2 *The Cubic Bezier Curve Segment*

The Bezier form of the parametric cubic is based on a different choice of basis functions, namely the cubic Bernstein polynomials. The curve segment is expressed as

$$\underline{r}(t) = (1-t)^3\underline{p}_o + 3t(1-t)^2\underline{p}_1 + 3t^2(1-t)\underline{p}_2 + t^3\underline{p}_3, \quad 0 \leqslant t \leqslant 1 . \tag{2.5}$$

In matrix form this becomes

$$\underline{r}(t) = [1 \;\; t \;\; t^2 \;\; t^3] \begin{bmatrix} 1 & 0 & 0 & 0 \\ -3 & 3 & 0 & 0 \\ 3 & -6 & 3 & 0 \\ -1 & 3 & -3 & 1 \end{bmatrix} \begin{bmatrix} \underline{p}_o \\ \underline{p}_1 \\ \underline{p}_2 \\ \underline{p}_3 \end{bmatrix}, \quad 0 \leqslant t \leqslant 1 . \tag{2.6}$$

The 'handles' or coefficients are all position vectors, and are related to the end points and tangents in the following way:

$$\begin{aligned} \underline{r}(0) &= \underline{p}_o & \dot{\underline{r}}(0) &= 3(\underline{p}_1 - \underline{p}_o) \\ \underline{r}(1) &= \underline{p}_3 & \dot{\underline{r}}(1) &= 3(\underline{p}_3 - \underline{p}_2) . \end{aligned} \tag{2.7}$$

From these relations it is apparent that, if an open polygon is defined by connecting the points $\underline{p}_o$, $\underline{p}_1$, $\underline{p}_2$, $\underline{p}_3$ in order by straight lines then the curve shares certain properties with the polygon. Not only does it have the same end points, but also it has the same end tangent directions, as illustrated in Fig. 2.2

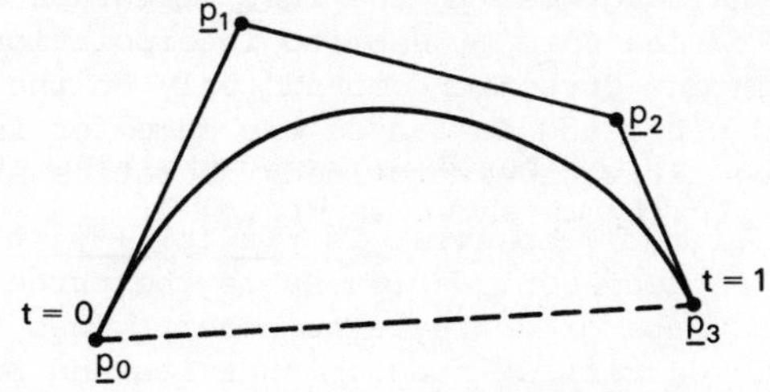

Fig. 2.2 Relation of Cubic Bezier Curve Segment to its Control Polygon

To be specific, the Bezier curve segment is a vector valued Bernstein approximation to the polygon. As pointed out by Davis [13], such approximations are in general qualitatively good but quantitatively bad. The first of these properties holds to the extent that manipulation of the polygon alters the curve in a predictable way. Accordingly the polygon is referred to as the control polygon for the curve and its vertices $\underline{p}_o$, $\underline{p}_1$, $\underline{p}_2$, $\underline{p}_3$ as control points.

The Bezier formulation has the virtue that the designer no longer has to think in terms of tangent vectors. Instead, he can define and manipulate his curves in an entirely intuitive way simply by specifying and moving control points.

The advantages of the Bezier formulation were first pointed out in the accessible literature by Bezier himself (Bezier [3,4]), though the principles had been independently established by de Casteljau [9] some years earlier.

2.3 The Cubic B-spline Curve Segment

Yet another choice of cubic basis functions gives rise to the cubic B-spline curve segment, which may be written as

$$\underline{r}(t) = \frac{1}{6}\begin{bmatrix}1 & t & t^2 & t^3\end{bmatrix}\begin{bmatrix}1 & 4 & 1 & 0\\ -3 & 0 & 3 & 0\\ 3 & -6 & 3 & 0\\ -1 & 3 & -3 & 1\end{bmatrix}\begin{bmatrix}\underline{p}_o\\ \underline{p}_1\\ \underline{p}_2\\ \underline{p}_3\end{bmatrix}, \qquad (2.8)$$

$$0 \leqslant t \leqslant 1 .$$

The $\underline{p}_i$ are position vectors of control points, as in the Bezier case, but the relation between the curve segment and the control points differs. From equation (2.8) it is easily seen that

$$\underline{r}(0) = \frac{2}{3}\underline{p}_1 + \frac{1}{6}\{\underline{p}_o + \underline{p}_2\}$$

and

$$\underline{r}(1) = \frac{2}{3}\underline{p}_2 + \frac{1}{6}\{\underline{p}_1 + \underline{p}_3\} .$$

The disposition of the curve segment with respect to its control polygon is typically as shown in Fig. 2.3.

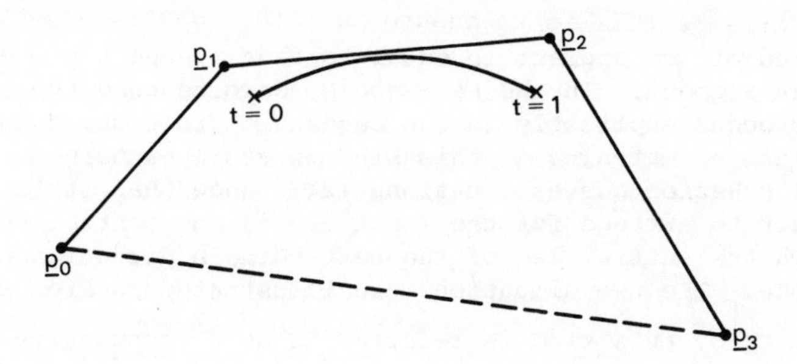

Fig. 2.3: Cubic B-spline Curve Segment and its Control Polygon

The B-spline curve formulation has many virtues, though these will not become apparent until composite curves are dealt with in Section 3. Further discussion is deferred until then, save for the remark that equation (2.8) represents only the simplest type of cubic B-spline curve segment; considerable generalisation is possible.

2.4 Convex Hull Property

For both the Bezier and the B-spline curve segments the cubic basis functions are all non-negative on [0,1] and sum to unity. Then any point on either type of segment is a weighted average of the four control points defining the associated polygon. This implies that the entire curve segment lies within the convex hull of its control points. Note that the convex hull of four coplanar points is the non self-intersecting quadrilateral with those points as vertices; if the points are non-coplanar their convex hull is a tetrahedron.

The property described has several important applications, notably in the calculation of geometric intersections (Pratt and Geisow [23], Sabin [26]).

2.5 Composite Cubic Curves

Single parametric cubic segments are not capable of representing the complex curves required in many engineering applications. However, such curves can be synthesised as sequences of cubic segments. In a practical situation the resulting composite curves must usually be tangent-continuous at the joins, and this is not difficult to achieve, as will be shown. For some applications curvature continuity at the joins is also desirable.

Consider first the case of tangent continuity. The obvious way to arrange this is to ensure that the tangent vector at the $t = 1$ end of one segment is equal to that at the $t = 0$ end of the next segment. In the Ferguson or Hermite case these vectors occur explicitly in the segment definition, equation (2.4), and so matching of this kind is easily attained. For composite Bezier curves equations (2.7) show that it is necessary to arrange for the final leg of one control polygon to match the initial leg of the next both in direction and magnitude. The two situations are illustrated in Fig. 2.4. Continuity of this kind is referred to as C^1 continuity; since $\dot{\underline{r}}(t)$ is continuous across the join in both direction and magnitude it is necessary for all three components $\dot{x}(t)$, $\dot{y}(t)$ and $\dot{z}(t)$ to be continuous there. As will be shown, this is an unnecessarily restrictive condition.

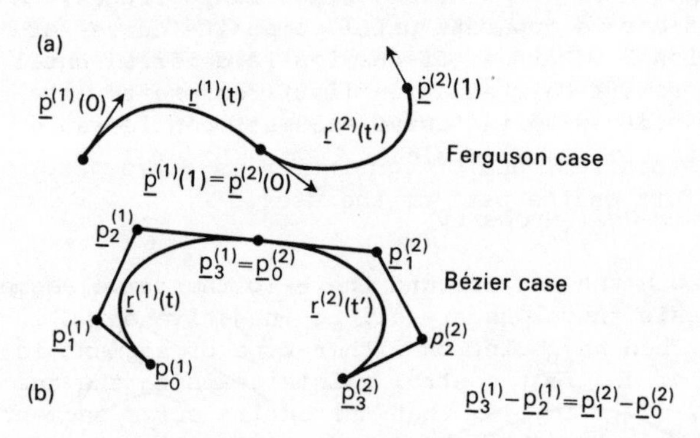

Fig. 2.4: C^1 Continuity between Cubic Curve Segments

In practice it is not necessary to match end tangents between segments in both direction and magnitude; if the matching is in direction only the curve is still geometrically smooth. In this case there will in general be discontinuities in $\dot{x}(t)$, $\dot{y}(t)$ and $\dot{z}(t)$ across the join, though their ratios must remain constant in order to preserve the directional continuity. This type of continuity allows more freedom in the construction of smooth composite curves. It has been referred to as G^1 continuity (Barsky and Beatty [2]), where the G stands for geometric.

It is also possible to obtain C^2 continuity for composite curves made up of cubic segments. For composite Hermite or Bezier curves the procedure involves a global computation in terms of an ordered set of points to be interpolated, whose result is appropriate values for tangent vectors at the data points. The composite curve has one cubic span between each successive pair of data points, and is referred to as a parametric cubic spline (Faux & Pratt) [14]).

Matters are much more convenient with the composite cubic B-spline curve, since C^2 continuity may be obtained automatically by arranging for successive control polygons to overlap as shown in Fig. 2.5. In this figure it can be seen that the segment $\underline{r}^{(1)}(t)$ is defined by the control points $\underline{p}_0$, $\underline{p}_1$, $\underline{p}_2$, $\underline{p}_3$, the segment $\underline{r}^{(2)}(t)$ is defined by $\underline{p}_1$, $\underline{p}_2$, $\underline{p}_3$, $\underline{p}_4$ and so on, each successive control polygon having three of its four points in common with the preceding one. An analysis based on equation (2.8) shows that $\underline{r}^{(1)}(t)$ at $t = 1$ and $\underline{r}^{(2)}(t)$ at $t = 0$ agree up to their second derivatives. This property of the composite B-spline curve is a very great virtue; as shown in Fig. 2.5. the entire composite curve may be defined in terms of a single polygon containing as many points as desired, and C^2 continuity is guaranteed with no extra effort on the part of the user.

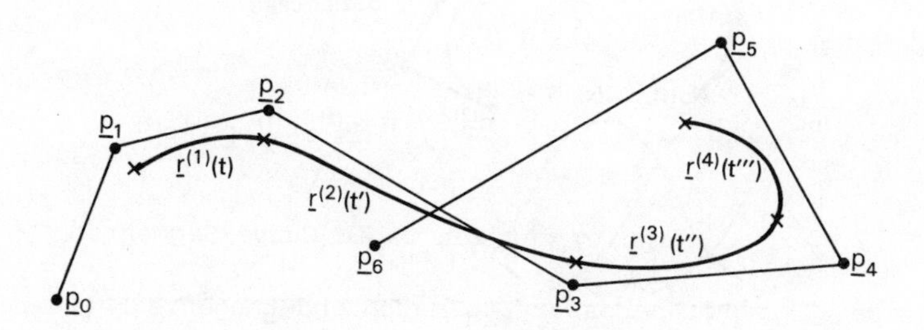

Fig. 2.5: Composite Cubic B-spline Curve

If a composite B-spline curve is defined entirely in terms of the curve segments of equation (2.8) it is said to be a uniform B-spline curve; the word 'uniform' denotes that all segments are uniformly parametrised from 0 to 1. It is also possible to construct B-spline curves for which this is not so, and the 'parametric length' of segments differs from one to the other. The resulting curve is said to be non-uniform (Gordon & Riesenfeld [20]), and there is now a considerable and rather complex literature on such curves (see, for example, Boehm [5], Cohen, Lyche & Riesenfeld [10]).

It was mentioned earlier that G^1 rather than C^1 continuity may be used to obtain additional flexibility in smooth composite curves, and it is possible to define G^2 continuity in a similar way. C^2 continuity requires $\underline{r}$, $\dot{\underline{r}}$ and $\ddot{\underline{r}}$ to be continuous across joins, which is very restrictive if the segments are of low degree. G^2 continuity results if G^1 continuity obtains and also the curvature vector is continuous across joins; compared with C^1 continuity this allows further extra degrees of freedom for the adjoining segments (Barsky and Beatty [2]).

2.6 Cubic B-splines

Expansion of the right-hand side of equation (2.8) reveals that the cubic B-spline curve segment on the interval $0 \leqslant t \leqslant 1$ is defined in terms of the set of basis functions $\frac{1}{6}(1-t)^3$, $\frac{1}{6}(4-6t^2+3t^3)$, $\frac{1}{6}(1+3t+3t^2-3t^3)$, $\frac{1}{6}t^3$. The curves representing these functions are depicted on the relevant interval in Fig. 2.6.

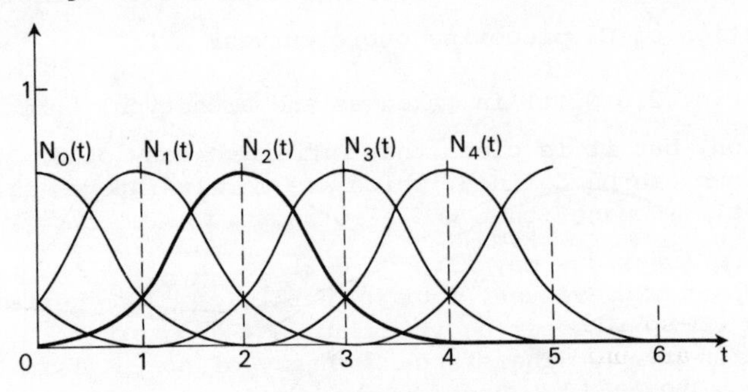

Fig. 2.6: Uniform Cubic B-Spline Functions

Now consider a composite B-spline curve defined by a set of control points $\underline{p}_0$, $\underline{p}_1$, ---, $\underline{p}_n$ as in Fig. 2.5. The second segment of the curve is expressed by equation (2.8) with the control points in the column vector replaced by $\underline{p}_1$, $\underline{p}_2$, $\underline{p}_3$, $\underline{p}_4$ and t replaced by t_2, a local parameter which runs from 0 to 1 on the new segment. However, if we set $t_2 = t - 1$ so that t runs from 1 to 2 on the second segment we see that we may use t as a single global parameter over both segments. Indeed, this stratagem allows parametrisation of the entire curve in terms of t, which runs from 0 to n over the n segments and is

related to the local parameter t_i on the ith segment by
$t_i = t - i + 1$.

In terms of the global parameter t the basis functions on the ith segment are simply those stated earlier for the interval $0 \leq t \leq 1$ translated a distance $i - 1$ along the t-axis. These are plotted for the first few intervals in Fig. 2.6, and it is immediately apparent from this figure that the basis functions on each integer interval in t are continuous with those on either adjacent interval. In fact an analysis based on equation (2.8) shows that C^2 continuity obtains between the curve segments where they join at integer values of t. Thus if we start at the origin we may follow the bell-shaped curve labelled $N_2(t)$; this C^2 curve is composed of four cubic segments. Further, the curve is C^2 continuous with the t - axis at $t = 0$ and $t = 4$, so that we have a C^2 continuous piecewise cubic function defined for all t but which departs from zero only on an interval of length precisely 4. This function is called a cubic B-spline; the B stands for basis because, as we shall see such functions may be used as a global basis for the definition of C^2 piecewise cubic curves.

In Fig. 2.6 $N_2(t)$ is taken as the archetypal cubic B-spline function, but it is clear that infinitely many other such functions can be defined, which are merely integer translates of $N_2(t)$. Some of them are also shown. As already mentioned, B-spline functions may also be defined on a non-uniformly spaced set of t-values, though in this case the members of the family will no longer be translates of each other but will differ in shape. The standard theory of non-uniform B-splines is given by Cox [12] and de Boor [8].

Now consider the curve defined by

$$\underline{r}(t) = \sum_{i=0}^{n} \underline{p}_i N_{i-1}(t), \quad 0 \leq t \leq n-2. \qquad (2.9)$$

If we restrict attention to the interval $0 \leq t \leq 1$ we find that only four terms contribute, since only $N_{-1}(t)$, $N_0(t)$, $N_1(t)$ and $N_2(t)$ are nonzero on this interval. Further, these four B-splines are identical on [0,1] with the four local basis cubics listed in the first paragraph of the present section. Then on this interval equation (2.9) reproduces the B-spline curve segment expressed by equation (2.8). Similarly, any other integer interval in t gives rise to another B-spline segment of the composite curve, which is known simply as a B-spline curve. Thus the set of cubics generated by equation (2.8), together with its integer translates, form local bases for the individual curve segments; the B-spline functions $N_i(t)$, however, provide a global basis for the composite curve as a whole.

B-spline curves possess a very important local modification property. Since the B-spline basis functions are each only nonzero over an interval of length 4, it is clear that altering any one of the control points $\underline{p}_i$ has only a local effect on the curve. This is a great virtue in curve design, which often requires geometric adjustment to be local in nature.

Various strategems have been devised for imposing end conditions of various types on B-spline curves. Some of these require the use of modified basis functions and others the use of 'phantom' control points. These matters are discussed in detail by Barsky [1].

2.7 The use of Higher Degree Polynomials

There is no difficulty in principle in extending the foregoing analysis in terms of polynomials of higher than cubic degree. Some examples follow:

1) Ferguson (Hermite) Curves: here we could go to degree 5 and fit $\underline{p}$, $\underline{\dot{p}}$ and $\underline{\ddot{p}}$ at each end of the curve segment. The disadvantage is that designers are not very happy to work in terms of first derivative vectors, let alone second derivatives.

2) Bezier Curves: here the curve can be based on Bernstein polynomials of degree $n > 3$, and defined in terms of n+1 control points. This preserves the desirable feature that the curve may be regarded as an approximation to its control polygon. The curve may therefore still be manipulated in terms of its control points in an intuitive manner, which explains why most of those practical systems which make use of higher degree polynomial bases are of Bezier type.

3) B-spline Curves: as in the Bezier case each segment may be defined in terms of a greater number of control points, though at the time of writing many workers have found that cubic B-spline curves are adequately flexible for most practical purposes.

The advantages of going to higher degree include greater flexibility and the possibility of higher degree continuity of composite curves. On the other hand the problem of presenting added degrees of freedom to the unsophisticated user in a way he can understand intuitively has not been solved. Furthermore G^1 continuity is sufficient for many practical purposes and there seems little point in seeking better than G^2 continuity, which can be achieved without going beyond cubics. Additionally, higher degree curves use more storage and require more complex algorithms. They are probably best reserved for use in special cases where curves have to meet stringent geometric constraints imposed by some desired relation to previously defined geometry.

3. PARAMETRIC CUBIC SURFACES

A parametric surface is defined in terms of two parameters, for which u and v will here be used. In the curve case it is usual to consider segments defined by values of t satisfying $0 \leqslant t \leqslant 1$, and the obvious extension is to restrict u and v to the unit square in parameter space, $0 \leqslant u \leqslant 1$, $0 \leqslant v \leqslant 1$. This gives a surface patch which is a mapping of the unit square into 3D, and which will in general have four curvilinear boundaries.

The most frequently used surface patches are those of the tensor - product type. These are defined in terms of basis functions having the form $\beta_i(u)\beta_j(v)$, where the β-functions are curve basis functions of the various kinds discussed earlier.

3.1 The Bicubic Ferguson Patch

The surface patch analogous to the Ferguson curve segment defined by equation (2.4) is given by

$$\underline{r}(u,v) = [1 \;\; u \;\; u^2 \;\; u^3]\; F\, Q\, F^T\, [1 \;\; v \;\; v^2 \;\; v^3]^T , \qquad (3.1)$$

$$0 \leqslant u \leqslant 1, \quad 0 \leqslant v \leqslant 1 ,$$

where

$$F = \begin{bmatrix} 1 & 0 & 0 & 0 \\ 0 & 0 & 1 & 0 \\ -3 & 3 & -2 & -1 \\ 2 & -2 & 1 & 1 \end{bmatrix} .$$

This is the matrix which, as previously, generates the Ferguson (Hermite) curve basis functions. The surface basis functions are sixteen in number; they comprise the set of all possible products of pairs of curve basis functions with one depending on u, the other on v. The central matrix Q contains the coefficients of these basis functions:

$$Q = \left[\begin{array}{cc|cc} \underline{p}(0,0) & \underline{p}(0,1) & \underline{p}_v(0,0) & \underline{p}_v(0,1) \\ \underline{p}(1,0) & \underline{p}(1,1) & \underline{p}_v(1,0) & \underline{p}_v(1,1) \\ \hline \underline{p}_u(0,0) & \underline{p}_u(0,1) & \underline{p}_{uv}(0,0) & \underline{p}_{uv}(0,1) \\ \underline{p}_u(1,0) & \underline{p}_u(1,1) & \underline{p}_{uv}(1,0) & \underline{p}_{uv}(1,1) \end{array}\right].$$

Here $\underline{p}_u$, $\underline{p}_v$ denote tangent vectors in the u and v directions respectively. Note that the patch is specified in terms of data specified at its four corners, as shown in Fig. 3.1.

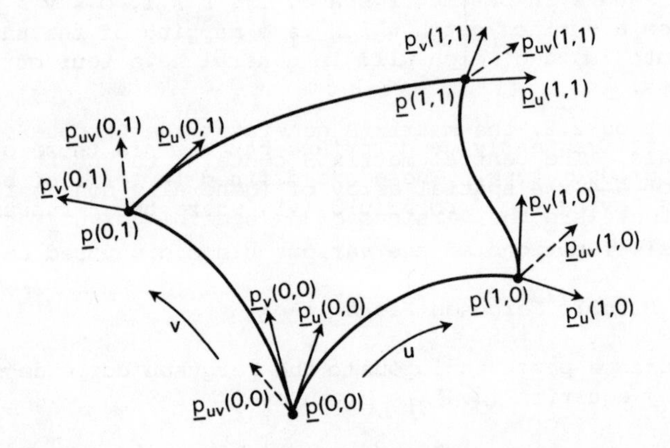

Fig. 3.1: A Ferguson Patch and its defining Data

The point and tangent vector information determine the boundary curves (which are standard curves of the Ferguson type). The $\underline{p}_{uv}$ are cross-derivative vectors which have effect only on the interior shape of the patch. They are sometimes referred to as <u>twist vectors</u>, but this is misleading terminology, since the values of $\underline{p}_{uv}$ are affected not only by the geometric twist in a surface, but also by the way in which the surface is parametrised (Faux & Pratt [14]). Most practical systems based on this type of surface provide automatic means for calculating acceptable values for the cross-derivatives since they do not have a readily comprehensible interpretation. In Ferguson's original formulation (Ferguson [15]) the twist vectors were all set to zero which leads to undesirable flattening near the patch corners.

3.2 The Bicubic Bezier Patch

By analogy with the analysis given in the last section we may write the bicubic Bezier patch as

$$\underline{r}(u,v) = [1 \quad u \quad u^2 \quad u^3]\; B\; S\; B^T\, [1 \quad v \quad v^2 \quad v^3]^T, \qquad (3.2)$$

$$0 \leqslant u \leqslant 1, \; 0 \leqslant v \leqslant 1,$$

where

$$B = \begin{bmatrix} 1 & 0 & 0 & 0 \\ -3 & 3 & 0 & 0 \\ 3 & -6 & 3 & 0 \\ -1 & 3 & -3 & 1 \end{bmatrix}$$

As in section 2.2, the matrix B generates the Bernstein basis polynomials. The central matrix S contains the position vectors of a 4 x 4 spatial array of (generally non-coplanar) points; these are the vertices of the control polyhedron of the surface patch:

$$S = \begin{bmatrix} \underline{p}_{00} & \underline{p}_{01} & \underline{p}_{02} & \underline{p}_{03} \\ \underline{p}_{10} & \underline{p}_{11} & \underline{p}_{12} & \underline{p}_{13} \\ \underline{p}_{20} & \underline{p}_{21} & \underline{p}_{22} & \underline{p}_{23} \\ \underline{p}_{30} & \underline{p}_{31} & \underline{p}_{32} & \underline{p}_{33} \end{bmatrix}.$$

The relation between the patch and the polyhedron is illustrated in Fig. 3.2.

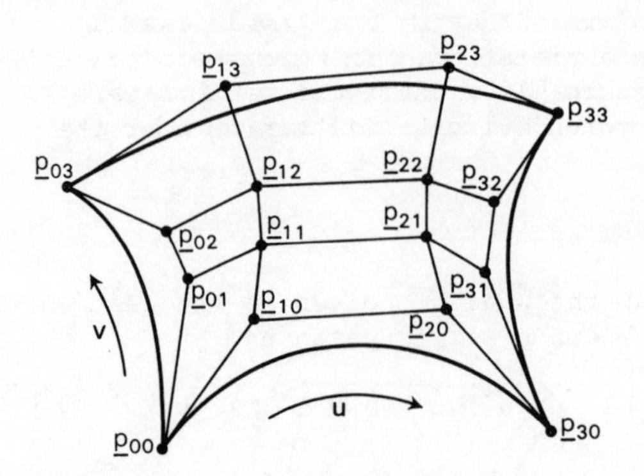

Fig. 3.2 A Bicubic Bezier Patch and its Control Polyhedron

In this patch definition the fact that only positional vector information is used is an advantage for the mathematically unsophisticated user. The four polygons forming the boundary of the control polyhedron are the control polygons of the patch boundary curves. The remaining four 'interior' polyhedron vertices control the interior shape of the patch, and are in fact related to the cross-derivative vectors at the patch corners referred to in the last section (Faux & Pratt [14]).

Like the Bezier curve, the Bezier surface has a convex hull property. It can be shown that the surface patch lies inside the convex hull of its control polyhedron.

3.3 The Bicubic B-spline Patch

Reasoning similarly, we define the bicubic B-spline patch by

$$\underline{r}(u,v) = [1 \quad u \quad u^2 \quad u^3]\, D\, T\, D^T\, [1 \quad v \quad v^2 \quad v^3]^T , \qquad (3.3)$$

$$0 \leq u \leq 1, \; 0 \leq v \leq 1 ,$$

where

$$D = \frac{1}{6}\begin{bmatrix} 1 & 4 & 1 & 0 \\ -3 & 0 & 3 & 0 \\ 3 & -6 & 3 & 0 \\ -1 & 3 & -3 & 1 \end{bmatrix}.$$

The matrix D was met earlier in equation (2.8) which defines a B-spline curve segment. As in the Bezier case the coefficient of the basis functions are the position vectors of a 4 x 4 array of space points which form the vertices of a control polyhedron. However, the relation between the patch and the polyhedron is different for the B-spline case, as shown in Fig. 3.3.

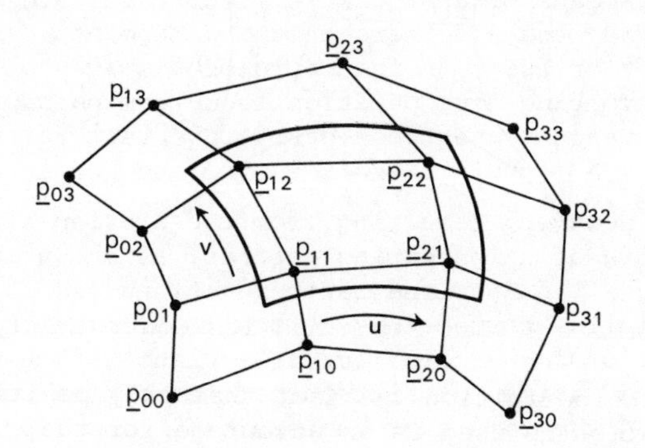

Fig. 3.3: A Bicubic B-spline Patch and its Control Polyhedron

The chief difference is that none of the control points lies on the surface in general, in contrast with the Bezier case where the corner points of the patch and the polyhedron coincide. The B-spline surface patch has the same convex hull property as was stated in the previous section for the Bezier patch.

3.4 Composite Bicubic Surfaces

It is easy to obtain C^1 continuity between two adjoining bicubic surface patches sharing a common boundary curve. Some examples are

(i) Ferguson patches - it is necessary and sufficient to match $\underline{p}$, $\underline{p}_u$, $\underline{p}_v$ and $\underline{p}_{uv}$ at the matching corners of the two patches.

(ii) Bicubic Bezier patches - the directions and lengths of polyhedron edges must be matched across the common polyhedron boundary defining the common boundary curve.

These conditions give C^1 continuity in all components of $\underline{r}(u,v)$ as functions of u and v across the boundary. Their use enables the building up of composite surfaces from many patches whilst retaining overall C^1 continuity. More general conditions may also be specified which lead to continuity of the G^1 type discussed in section 2.5 (Hagen [22]).

If overall C^2 continuity is required of a composite surface then a suitable quasi-rectangular array of space points may be interpolated by means of a piecewise bicubic surface made up of patches having their corners at the datapoints. In the Ferguson and Bezier cases an initial global survey of the data is required, involving a computation to determine appropriate values for the tangent and cross-derivative vectors at the patch corners. A bicubic B-spline surface has automatic C^2 continuity, however, resulting from the overlap of the control polyhedra of neighbouring patches, as shown in Fig. 3.4. Moreover, because the B-spline basis functions are only locally nonzero the B-spline surface has a local modification property similar to that of the B-spline curve; a change in one control point modifies a maximum of sixteen patches of the composite surface, leaving the remainder undisturbed. These properties of B-spline surfaces were first expounded by Gordon and Riesenfeld [20].

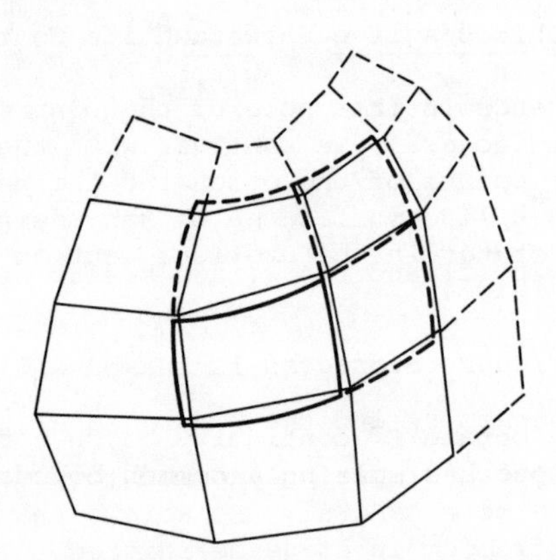

Fig. 3.4: Part of the Control Polyhedron of a Bicubic B-spline Surface, showing overlap of the 4 x 4 Polyhedron controlling neighbouring Patches

As in the curve case, it is possible to obtain additional freedom in the definition of composite surfaces by working in terms of G^2 rather than C^2 continuity (Hagen [22]).

4. SOME EXTENSIONS OF THE THEORY

There are certain advantages to be gained from defining curves and surfaces in terms of functions more general than polynomials. In this last section we first briefly consider the use of rational functions and finally give an outline of the theory of Coons surfaces, which allows the construction of surface patches in very general terms.

4.1 Rational Polynomial Curves and Surfaces

It has been shown that a parametric polynomial curve of degree n can be written in terms of its basis functions $\beta_i(t)$ as

$$\underline{r}(t) = \sum_{i=o}^{n} \underline{c}_i \beta_i(t).$$

More generally, a rational curve may be defined by

$$\underline{r}(t) = \frac{\sum_{i=o}^{n} \underline{c}_i w_i \beta_i(t)}{\sum_{i=o}^{n} w_i \beta_i(t)} = \frac{1}{w(t)} \sum_{i=o}^{n} \underline{c}_i w_i \beta_i(t) ,$$

where $w(t) = \sum_{i=o}^{n} w_i \beta_i(t)$ and the w_i are scalar weights. Note that if the $\beta_i(t)$ sum to one (as is the case with Bezier and B-spline curve segments) and the w_i are equal, the above equation reduces to a simple polynomial form. The rational form is therefore more general; the weights may be manipulated to gain greater freedom in curve definition.

In fact the rational form shown can be re-expressed as a polynomial form provided we work in terms of four-dimensional vectors using homogeneous coordinates (Riesenfeld [24]). On multiplying the last equation through by w(t) and writing ξ_i, η_i, ζ_i for the components of $\underline{c}_i$ we obtain the component equations

$$w(t)x(t) = \sum \xi_i \, w_i \, \beta_i(t)$$
$$w(t)y(t) = \sum \eta_i \, w_i \, \beta_i(t)$$
$$w(t)z(t) = \sum \zeta_i \, w_i \, \beta_i(t)$$

When taken together with the additional equation

$$w(t) = \Sigma\, w_i\, \beta_i(t)$$

these equations may be summarised as

$$\underline{R}(t) = \sum_{i=o}^{n} \underline{C}_i \beta_i(t) \; .$$

in which $\underline{R} = (wx, wy, wz, w)^T$ and $\underline{C}_i = (w_i\xi_i, w_i\eta_i, w_i\zeta_i, w_i)^T$.

There are several advantages to the use of rational forms:

(i) It is possible to modify a curve specified by a given set of defining data by adjustment of the weights.

(ii) It is also possible to adjust the weights so that the geometry of the curve is unaffected but its parametrisation is changed.

(iii) (Perhaps most important) The use of rational forms allows the exact representation of conic curves, which may always be expressed as parametric rational quadratics. Many industries, notably the aircraft industry, have traditionally used conic curves in design, and have found it a disadvantage of the purely polynomial-based system that it can only represent conics approximately.

The theory of rational biparametric surfaces requires a similar extension to that outlined in Section 3 of this paper. Further discussion is given in Faux and Pratt [14].

4.2 Coons Patches

In the early sixties Coons devised a powerful method for interpolating a surface patch to fit four prescribed boundary curves, with no restriction on their type, subject only to the condition that they form a continuous closed boundary to the patch, as shown in Fig. 4.1.

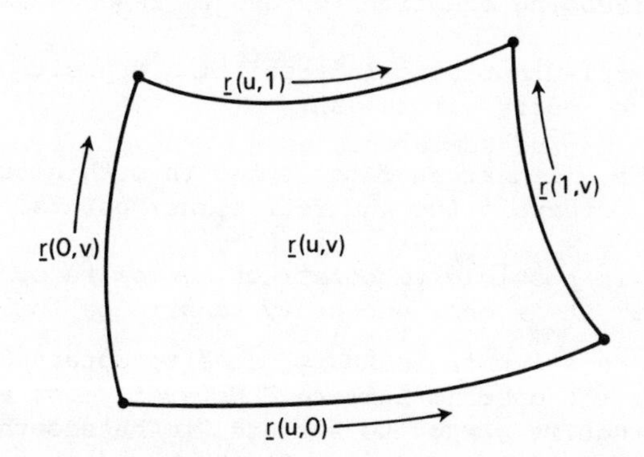

Fig. 4.1: Coons Patch fitting four prescribed Boundary Curves

Suppose the four boundary curves are $\underline{r}(u,0)$, $\underline{r}(u,1,)$ $\underline{r}(0,v)$ and $\underline{r}(1,v)$ as shown. Coons showed that the simplest surface fitting the boundary data is expressible in terms of two <u>blending functions</u> $\alpha_0(*)$ and $\alpha_1(*)$, which must be monotone on the interval $[0,1]$ and possess the following additional properties:

$$\alpha_0(0) = 1 \quad \alpha_0(1) = 0$$

$$\alpha_1(0) = 0 \quad \alpha_1(1) = 1$$

$$\alpha_0(*) + \alpha_1(*) = 1$$

A solution to Coons' interpolation problem is then

$$\underline{r}(u,v) = [\alpha_0(u) \ \alpha_1(u)] \begin{bmatrix} \underline{r}(0,v) \\ \underline{r}(1,v) \end{bmatrix} + [\underline{r}(u,0) \ \underline{r}(u,1]] \begin{bmatrix} \alpha_0(v) \\ \alpha_1(v) \end{bmatrix}$$

$$- [\alpha_0(u) \ \alpha_1(u)] \begin{bmatrix} \underline{r}(0,0) & \underline{r}(0,1) \\ \underline{r}(1,0) & \underline{r}(1,1) \end{bmatrix} \begin{bmatrix} \alpha_0(v) \\ \alpha_1(v) \end{bmatrix}, \qquad (4.1)$$

$$0 \leqslant u \leqslant 1, \quad 0 \leqslant v \leqslant 1.$$

Substitution of $u = 0,1$ and $v = 0, 1$ quickly shows this surface to have the required interpolation properties. This type of interpolation is sometimes called <u>transfinite</u> (Gordon [19]) because it goes beyond the classical interpolation techniques which only interpolate conditions at a finite number of points.

It should be noted that the three terms on the right-hand side of the foregoing equation represent, respectively,

(i) and (ii) ruled surfaces interpolating opposite pairs of boundary curves, and

(iii) a bilinear surface, ruled in both directions, interpolating the four corner points.

Clearly it is possible to construct composite surfaces from Coons patches. It is only necessary to arrange for adjacent patches to share a common boundary curve to obtain C^0 continuity for the overall surface. However, most engineering applications require composite surfaces to be smooth. To achieve this we may specify not only the boundary curves but also the cross-boundary tangent vector distributions, which may be denoted by $\underline{r}_v(u,0)$, $\underline{r}_v(u,1)$, $\underline{r}_u(0,v)$ and $\underline{r}_u(1,v)$. To interpolate this extra information two new blending functions $\beta_0(*)$ and $\beta_1(*)$ are needed, and additional restrictions must be imposed upon $\alpha_0(*)$ and $\alpha_1(*)$. We now require

$$\alpha_0'(0) = \alpha_0'(1) = \alpha_1'(0) = \alpha_1'(1) = 0$$

$$\beta_0(0) = \beta_0(1) = \beta_1(0) = \beta_1(1) = 0$$

$$\beta_0'(0) = 1 \ , \ \beta_0'(1) = 0$$

$$\beta_1'(0) = 0 \ , \ \beta_1'(1) = 1 \ .$$

The resulting Coons patch is given by

$$\underline{r}(u,v) = [\alpha_0(u)\ \alpha_1(u)\ \beta_0(u)\ \beta_1(u)] \begin{bmatrix} \underline{r}(0,v) \\ \underline{r}(1,v) \\ \underline{r}_u(0,v) \\ \underline{r}_u(1,v) \end{bmatrix}$$

$$+ [\underline{r}(u,0)\ \underline{r}(u,1)\ \underline{r}_v(u,0)\ \underline{r}_v(u,1)] \begin{bmatrix} \alpha_0(u) \\ \alpha_1(u) \\ \beta_0(u) \\ \beta_1(u) \end{bmatrix} \qquad (4.2)$$

$$- [\alpha_0(u)\alpha_1(u)\beta_0(u)\beta_1(u)] \begin{bmatrix} \underline{r}(0,0) & \underline{r}(0,1) & \underline{r}_v(0,0) & \underline{r}_v(0,1) \\ \underline{r}(1,0) & \underline{r}(1,1) & \underline{r}_v(1,0) & \underline{r}_v(1,1) \\ \underline{r}_u(0,0) & \underline{r}_u(0,1) & \underline{r}_{uv}(0,0) & \underline{r}_{uv}(0,1) \\ \underline{r}_u(1,0) & \underline{r}_u(1,1) & \underline{r}_{uv}(1,0) & \underline{r}_{uv}(1,1) \end{bmatrix} \begin{bmatrix} \alpha_0(v) \\ \alpha_1(v) \\ \beta_0(v) \\ \beta_1(v) \end{bmatrix}$$

Note that once again the first two terms on the right-hand side involve data specified on opposite pairs of boundaries while the third term involves only patch corner data. A C^1 composite surface may now be constructed by sharing boundary curves and cross-boundary tangent functions between adjacent patches.

Coon's original work was published in a report from Massachusetts Institute of Technology (Coons [11]). More recent discussions of his approach are to be found in Forrest [17] and Faux & Pratt [14].

In practice, Coons patches are not often used in their full generality. The simpler form (4.1) is used in some commercial systems, usually with the obvious choice of blending functions

$$\alpha_o(t) = 1 - t \ , \ \alpha_1(t) = t.$$

In the case of equation (4.2) the simplest set of polynomial blending functions satisfying all the required conditions is

$$\alpha_o(t) = 1 - 3t^2 + 2t^3, \quad \alpha_1(t) = 3t^2 - 2t^3$$

$$\beta_o(t) = t - 2t^2 + t^3, \quad \beta_1(t) = -t^2 + t^3 \ .$$

But this is precisely the set of Hermite basis functions first introduced in equation (2.3), in the context of Ferguson cubic curves. In fact, if all the boundary curves and cross-boundary tangent conditions are now specified in the Ferguson manner, in terms of corner data and Hermite cubic blending functions, then equation (4.2) simplifies considerably. All three terms on the right-hand side become identical, and consequently two of them cancel out; the remaining term proves to be the same as the right-hand side of equation (3.1), so that the Coons patch reduces to the frequently used Ferguson bicubic patch for this particular case. Unfortunately equation (3.1) is often erroneously regarded as the fundamental definition of a Coons patch whereas, as has been shown, it is merely one rather simple instance of such a patch.

4.3 Composite Coons Surfaces

A straightforward extension of the theory of the last section permits the interpolation of a composite surface to fit a mesh defined by two families of intersecting curves as shown in Fig. 4.2.

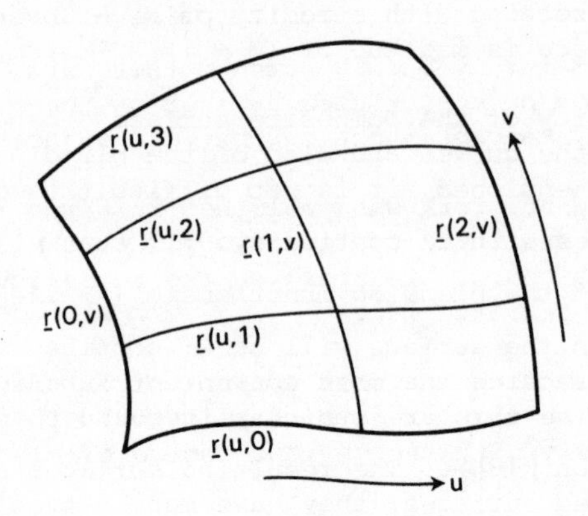

Fig. 4.2: Curve Mesh to be Interpolated by a Composite Coons Surface

The curves forming the mesh may be regarded as constant parameter lines $\underline{r}(u,o)$, $\underline{r}(u,1)$, $\underline{r}(u,2)$,..... and $\underline{r}(o,v)$, $\underline{r}(1,v)$, $\underline{r}(2,v)$, ... of the overall surface. The simplest case is when there are just two curves in each direction; a single patch may then be defined by equation (4.1) stated earlier, which may be rewritten in the form

$$\underline{r}(u,v) = \sum_{i=o}^{1} \alpha_i(u)\underline{r}(i,v) + \sum_{j=o}^{1} \alpha_j(v)\underline{r}(u,j) - \sum_{i=o}^{1}\sum_{j=o}^{1} \alpha_i(u)\alpha_j(v)\underline{r}(i,j),$$

in which $\alpha_k(\ell) = \delta_{k\ell}$, $k, \ell \in \{O,1\}$.

This permits the ready generalisation

$$\underline{r}(u,v) = \sum_{i=o}^{M} \alpha_i(u)\underline{r}(i,v) + \sum_{j=o}^{N} \alpha_j(v)\underline{r}(u,j) - \sum_{i=o}^{M}\sum_{j=o}^{N} \alpha_i(u)\alpha_j(v)\underline{r}(i,j), \quad (4.3)$$

where $\alpha_k(\ell) = \delta_{k\ell}$, $k, \ell \in \{O,1,2,\ldots, \max(M,N)\}$.

This surface interpolates a mesh defined by a family of M v-curves intersecting with a family of N u-curves; the resulting surface is made up of (M - 1)(N - 1) patches.

The continuity of the composite surface depends on the continuity of the curves and also of the blending functions. As previously mentioned, it is not difficult to define composite curves with C^2 continuity. If the mesh is made up of such curves and the blending functions $\alpha_k(*)$ are also C^2 continuous then the surface will be C^2 continuous between patches. In practice the most convenient functions to use for the α_k to achieve this are cubic splines known as cardinal splines (Gordon [18]). The resulting surfaces are called spline - blended surfaces; they have many attractive properties but are expensive to compute.

4.4 Non Four-Sided Surface Patches

There are various practical situations where it is desirable to use surface patches with three or five sides (sometimes even more). There are several approaches which allow these to be defined. Firstly there are triangular patch formulations of the Bezier type, originally due to de Casteljau (de Casteljau [9]). Secondly there are methods based on the use of projectors, which provide an elegant alternative means for deriving the Coons patch formulae (4.1) and (4.2) but are capable of considerable generalisation (Gordon [19], Gregory [21]). Thirdly there is an emerging theory of multivariate B-splines (Boehm [6]). A great deal of fruitful research is currently in progress in this general area, but most of the results still await application in practical systems. In the meantime most surface definition software is based on the four-sided patch formulations discussed earlier in this survey; if three-sided patches are required they are generally defined as 'degenerate' four-sided patches in which the length of one boundary curve is zero. Such patches often give rise to problems because care is needed to ensure that a well-defined surface normal exists at the degenerate corner.

5. CONCLUSIONS

In this introductory survey it has only been possible to outline the bare essentials of the subject. The references provided are intended to point the way to fuller expositions of the individual topics covered. Boehm, Farin and Kahmann [7] have recently published a more extended survey which is in many ways complementary to this paper; this is recommended further reading since it contains references to much recent work in the area. Two introductory books dealing with much of the material of this paper in further detail are Rogers and Adams [25] and Faux and Pratt [14].

REFERENCES

[1] Barsky, B.A., End Conditions and Boundary Conditions for Uniform B-spline Curve and Surface Representations, Computers in Industry **3**, pp 17 - 29, 1982.

[2] Barsky, B.A. & Beatty, J.C., Local Control of Bias and Tension in Beta-Splines, ACM Transactions on Graphics **2**, pp 109 - 134, 1983.

[3] Bezier, P., *Numerical Control : Mathematics and Applications*, Wiley, 1972.

[4] Bezier, P., Mathematical and Practical Possibilities of UNISURF, in *Computer Aided Geometric Design*, eds. R.E. Barnhill & R.F. Riesenfeld, Academic Press, 1974.

[5] Boehm, W., Inserting New Knots into B-spline Curves, Computer Aided Design **12**, pp 199 - 201, 1980.

[6] Boehm, W., Multivariate Spline Algorithms, this volume.

[7] Boehm, W, Farin, G. & Kahmann, J., A Survey of Curve and Surface Methods in CAGD, Computer Aided Geometric Design **1**, pp 1 - 60, 1984.

[8] de Boor, C., On Calculating with B-splines, J Approx Th. **6**, pp 50 - 62, 1972.

[9] de Casteljau, F., Courbes et Surfaces a Poles, Andre Citroen Automobiles S.A., Paris, 1962.

[10] Cohen, E., Lyche, T. & Riesenfeld, R.F., Discrete B-splines and Subdivision Techniques in Computer Aided Geometric Design and Computer Graphics, Computer Graphics and Image Processing **14**, pp 87 - 111, 1980.

[11] Coons, S.A., Surfaces for Computer Aided Design of Space Forms, Report MAC-TR-41, Project MAC, MIT, 1967.

[12] Cox, M.G., The Numerical Evaluation of B-splines, J. Inst. Maths Applics. **10** pp 134 - 149, 1972.

[13] Davis, P.J., *Interpolation and Approximation*, Dover, 1963.

[14] Faux, I.D. & Pratt, M.J., *Computational Geometry for Design and Manufacture*, Ellis Horwood, 1979.

[15] Ferguson, J.C., Multivariable Curve Interpolation, J.ACM **11**, pp 221 - 228, 1964.

[16] Forrest, A.R., Mathematical Principles for Curve and Surface Representation, in *Curved Surfaces in Engineering*, ed. J. Brown, IPC Science & Technology Press, 1972.

[17] Forrest, A.R., On Coons' and other methods for the Representation of Curved Surfaces, Computer Graphics and Image Processing 1, pp 341 - 359, 1972.

[18] Gordon, W.J., Spline-blended Surface Interpolation through Curve Networks, J. Mathematics & Mechanics **18**, pp 931 - 952, 1969.

[19] Gordon, W.J., An Operator Calculus for Surface and Volume Modelling, IEEE Computer Graphics & Applications **3** pp 18 - 22, 1983.

[20] Gordon, W.J. & Riesenfeld, R.F., B-spline Curves and Surfaces, in *Computer Aided Geometric Design*, eds. R.E. Barnhill & R.F. Riesenfeld, Academic Press, 1974.

[21] Gregory, J.A., N-sided Patches, this volume.

[22] Hagen, H., Surfaces with Geometric Continuity, Proc. Conf. on *Surfaces in Computer Aided Geometric Design*, Oberwolfach, Germany, 12 - 16 November 1984, North-Holland, (to be published).

[23] Pratt, M.J. & Geisow, A.D., Surface/Surface Intersection Problems, this volume.

[24] Riesenfeld, R.F., Homogeneous Coordinates and Projective Planes in Computer Graphics, IEEE Computer Graphics & Applications, **1** pp 50 - 55, 1981.

[25] Rogers, D.F. & Adams, J.A., *Mathematical Elements for Computer Graphics* McGraw-Hill, 1976.

[26] Sabin, M.A., Recursive Division Techniques, this volume.

IMPROVEMENTS TO PARAMETRIC BICUBIC SURFACE PATCHES

D.P. Sturge
(Delta C.A.E. Ltd., Cambridge)

1. INTRODUCTION

When devising a surface patch system for use in engineering design it is important to provide handles with which to control the shape, but not so many as to confuse the non-mathematical designer. The starting point for this paper is a network of lines enclosing four-sided regions called patches, as in Figure 1. The problem is then how to generate the surface patches that the designer intuitively expects, for a given network of lines. The problem of defining the lines themselves, by splining or otherwise estimating slopes and tangent magnitudes, is not treated here. The lines are assumed to be parametric cubic curves, as described in Chapter 5 of Faux and Pratt [1].

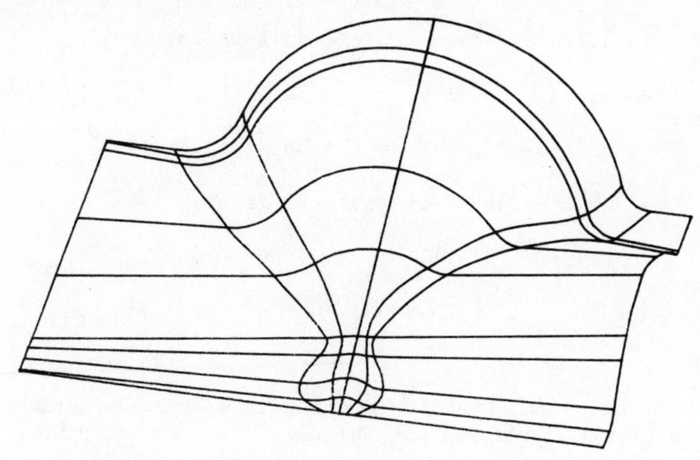

Fig. 1 Network of Lines

Surface patches which clothe a network of lines should meet the following conditions, as far as possible:

a) The surface normal shall be continuous over the interior of each patch, and across all patch boundaries.

b) The interior of a patch shall be as well conditioned as its boundaries permit. It shall be fair, without unexpected inflexions, concentrations of curvature, folds or loops.

c) The shape of each patch shall be independent of the shapes of adjacent patches.

d) A region of a surface which has boundaries consistent with a surface of revolution shall represent such a surface, to the same accuracy that a parametric cubic curve represents a circular arc.

e) The surface shall be generated automatically, without the explicit specification of any internal control points, shape factors or cross-derivative vectors.

When the patch boundaries are all parametric cubic curves it is obvious to use a Cartesian product bicubic surface. The Bezier representation is used here, with its four interior control points in each patch, as in Figure 2. A point $\underline{r}$ on such a bicubic surface is then given by

$$\begin{aligned}
\underline{r}(u,v) = {} & (1-u)^3 . (1-v)^3 \underline{p}_{00} + (1-u)^3 .3v(1-v)^2 \underline{a}_{00} \\
& + (1-u)^3 .3v^2(1-v) \underline{a}_{01} + (1-u)^3 . v^3 \underline{p}_{01} \\
& + 3u(1-u)^2 . (1-v)^3 \underline{b}_{00} + 3u(1-u)^2 .3v(1-v)^2 \underline{i}_{00} \\
& + 3u(1-u)^2 .3v^2(1-v) \underline{i}_{01} + 3u(1-u)^2 . v^3 \underline{b}_{01} \\
& + 3u^2(1-u). (1-v)^3 \underline{b}_{10} + 3u^2(1-u).3v(1-v)^2 \underline{i}_{10} \\
& + 3u^2(1-u).3v^2(1-v) \underline{i}_{11} + 3u^2(1-u). v^3 \underline{b}_{11} \\
& + u^3 . (1-v)^3 \underline{p}_{10} + u^3 .3v(1-v)^2 \underline{a}_{10} \\
& + u^3 .3v^2(1-v) \underline{a}_{11} + u^3 . v^3 \underline{p}_{11}
\end{aligned} \qquad (1.1)$$

There are problems in obtaining continuity of surface normal across the patch boundaries with bicubic surfaces, without sacrificing the independence of the boundaries. These limitations are discussed in Section 2.

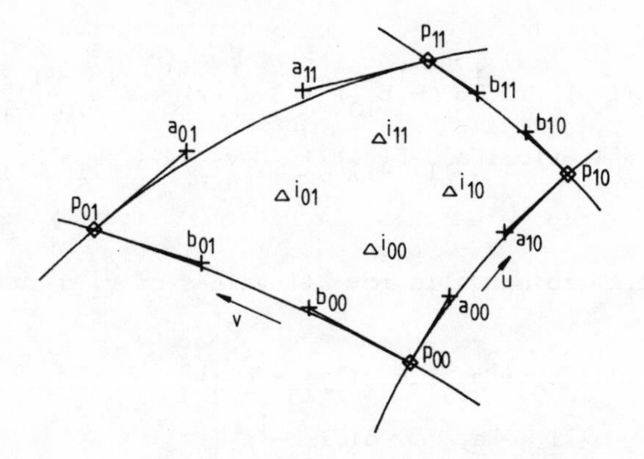

Fig. 2 Bezier Control Points

Section 3 describes three alternative ways of obtaining surface normal continuity with independent boundaries. All of these ways involve raising the order of the patches.

Section 4 describes a method of selecting positions for the interior control points and the associated corner cross-derivatives automatically. The method satisfies requirements (b) and (d) above for internal fairness as far as possible. Two components of the corner cross-derivatives can be made to satisfy requirement (c), but not the third component.

2. CONTINUITY WITH BICUBIC PATCHES

In this section the restrictions on the layout of data and interior points for continuity of surface normal across the boundary between two bicubic patches are formulated. Consider the tangent vector of $\underline{r}$ on two sides of a boundary of constant u between two patches 1 and 2 as in Figure 3 and denote the two patches by ${}^{1}\underline{r}$ and ${}^{2}\underline{r}$. For normal continuity at any v along this boundary it is sufficient that the tangent vector is continuous in direction but not magnitude across the boundary. Thus

$$ {}^{2}\underline{r}_{u}(0,v) = g\,{}^{1}\underline{r}_{u}(1,v) \qquad (2.1) $$

where g is a scalar function of v.
Now

$$ \begin{aligned} {}^{2}\underline{r}_{u}(0,v) = {} & (1-v)^{3}\,.3({}^{2}\underline{b}_{00}-{}^{2}\underline{p}_{00}) + 3v(1-v)^{2}.3({}^{2}\underline{i}_{00}-{}^{2}\underline{a}_{00}) \\ & + 3v^{2}(1-v).3({}^{2}\underline{i}_{01}-{}^{2}\underline{a}_{01}) + v^{3}\,.3({}^{2}\underline{b}_{01}-{}^{2}\underline{p}_{01}) \end{aligned} \qquad (2.2) $$

and

$$ {}^1\underline{r}_u(1,v) = (1-v)^3\,.3({}^1\underline{p}_{10}-{}^1\underline{b}_{10}) + 3v(1-v)^2.3({}^1\underline{a}_{10}-{}^1\underline{i}_{10}) $$
$$ + 3v^2(1-v).3({}^1\underline{a}_{11}-{}^1\underline{i}_{11}) + v^3\,.3({}^1\underline{p}_{11}-{}^1\underline{b}_{11}) \qquad (2.3) $$

For equation (2.1) to be valid for all values of v, g must be a constant,

$$ ({}^2\underline{b}_{0j}-{}^2\underline{p}_{0j}) = g({}^1\underline{p}_{1j}-{}^1\underline{b}_{1j}) $$

and

$$ ({}^2\underline{i}_{0j}-{}^2\underline{a}_{0j}) = g({}^1\underline{i}_{1j}-{}^1\underline{a}_{1j}) \qquad (2.4) $$

for j = 0,1, from comparison of coefficients of powers of v.

The first consequence of (2.4) is that the ratios of tangent magnitudes before and after each data point must be the same as one moves from one patch boundary to the next. This requirement is incompatible with the independence of all the patch boundaries.

The second consequence is that lines of the type $\underline{i}$ - $\underline{a}$ - $\underline{i}$ shown dashed in Figure 3 must be straight, and that the same ratio g must apply on these lines. Similar considerations of continuity across lines of the other family require that lines of the type $\underline{i}$ - $\underline{b}$ - $\underline{i}$ must also be straight with corresponding ratios of tangent magnitudes, imposing strict constraints on the positioning of the interior control points $\underline{i}$.

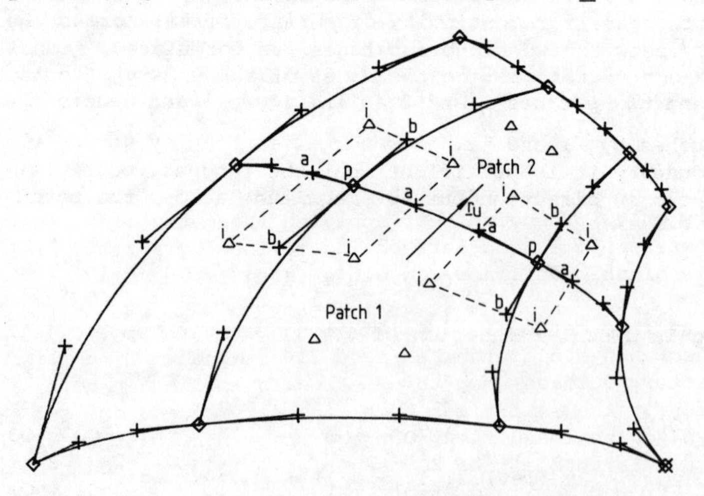

Fig. 3 Continuity across Boundaries

An alternative approach is not to enforce continuity of tangent direction as in equation (2.1), but rather to insist that the tangents remain coplanar, by specifying

$$^{2}\underline{r}_u(0,v) = g\,^{1}\underline{r}_u(1,v) + h\,^{1}\underline{r}_v(1,v) \tag{2.5}$$

Substitution of expressions for $\underline{r}_u$ and $\underline{r}_v$ and comparison of coefficients of powers of v shows that g is constant and h is a linear function of v. Since we require that h = 0 at v = 0 and 1 for continuity of patch boundary curves, this alternative approach is not helpful for bicubic patches. With higher order patches this approach can be used to obtain continuity, as in section 3.3.

3. HIGHER ORDER PATCHES

In this section three methods of overcoming the problems of fitting patches to parametric cubic boundaries are described. Each method involves raising the order of the patches to a greater or lesser extent.

3.1 Independent blending of magnitude and direction

In Section 2 the requirement for a constant ratio of tangent magnitude across patch boundaries was described. A way of meeting this requirement is to separate the vectors $\underline{b} - \underline{p}$ and $\underline{i} - \underline{a}$ in Figure 3 into magnitude and unit vectors. The vector $^{2}\underline{r}_u(0,v)$ is then found by applying the usual cubic blending to both the magnitudes and the unit vectors separately, and recombining them. A point $\underline{r}(u,v)$ within a patch may then be found by again applying the usual cubic blending in the u direction to $\underline{r}(0,v)$, $\underline{r}_u(0,v)$, $\underline{r}_v(1,v)$ and $\underline{r}(1,v)$, as in Figure 4.

As just described, the method is applied to one type of boundary only. The effect of applying the method in this way is to generate a lofted surface (as defined by Forrest [2]) which is ninth order in v and cubic in u.

If the same method is applied along boundaries of constant v instead then a different surface is obtained. To obtain a satisfactory surface when this method is applied in both direction at the same time it is necessary to add the two surfaces together and then subtract the simple bicubic surface. The result is then a true Coons surface, as described by Forrest [2] and Faux and Pratt [1].

The advantages of this method of fitting a surface are that tangent vectors are always continuous in direction across patch boundaries, however irregular, and the only limitation on the positions of interior points is that lines $\underline{i} - \underline{a} - \underline{i}$ and $\underline{i} - \underline{b} - \underline{i}$ in Figure 3 must be straight.

The disadvantages are that the surface is ninth order, and that each surface point requires a relatively large amount of calculation.

3.2 Redistribution of parameter

The second approach considered here is to reparametrise the boundaries before creating the surface. In order that the ratio g should be constant in equations 2.1 and 2.4, we can reassign the parameter distribution along the boundaries of constant v. Suppose that the boundary curves are ordinary parametric cubics defined in terms of a parameter t, and the surface is defined in terms of u. We wish to specify du/dt at the ends, and choose a suitable function relating t and u with the following properties:

At $t = 0$, $u = 0$ and du/dt is defined.
At $t = 1$, $u = 1$ and du/dt is defined.
For all $0 \leq t \leq 1$, $du/dt > 0$. (3.1)

Several functions can be used, but a rational quadratic has been found to have good properties from a practical point of view.

It is necessary to assign values of du/dt to both ends of all the lines of constant v, such that the ratio g is made constant across all lines of constant u. It is desirable to keep the values of du/dt as close to 1 as possible, to keep the surface smooth.

The procedure for finding a point $\underline{r}(u,v)$ on the surface is then as follows:

a) find the values of t on each boundary from u.

b) with these values of t, find $\underline{r}$ and $\underline{r}_v$ along each side by ordinary cubic blending.

c) Find $\underline{r}(u,v)$ from $\underline{r}$ and $\underline{r}_v$ on the two sides by ordinary cubic blending, in a similar manner to that in Figure 4 but with the other parameter.

As with the method of 3.1, the surface is a lofted surface when applied to one type of boundary only. When applied to both types of boundary the full Coons must be used to calculate points.

Compared with the method of 3.1 the complexity of calculation of surface points is about the same. If however redistribution of parameter is used on one type of boundary and separate blending of magnitudes and directions on the other, then the resulting surface is only lofted as in Figure 4, and not a Coons surface. This mixed approach is adopted in the commercially available DUCT system described by Sturge and Nicholls [3] and Welbourn [4].

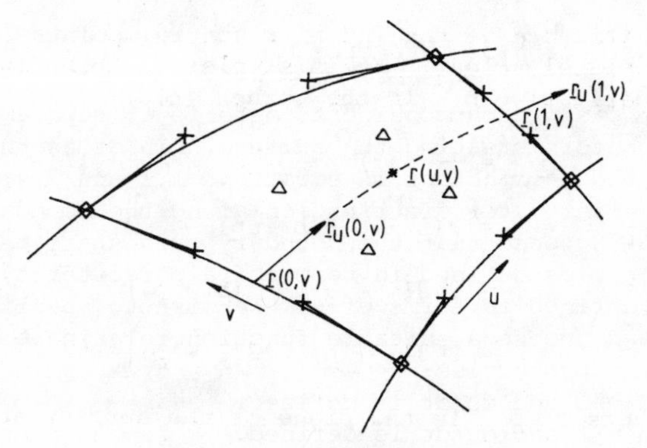

Fig. 4 Lofted Surface Patch

3.3 Alternate patches of higher order

In Section 2, equation (2.5), a method of obtaining continuity of surface normal but not of tangent direction was outlined as

$$^2\underline{r}_u(0,v) = g\,^1\underline{r}_u(1,v) + h\,^1\underline{r}_v(1,v)$$

If patch 1 is bicubic, and patch 2 is fourth order, then g and h can be useful functions of v.

$$g = g_0(1-v) + g_1 v$$

$$h = h_0 v(1-v) \tag{3.2}$$

so that at $v = 0$ $g = g_0$ and $h = 0$, and

at $v = 1$ $g = g_1$ and $h = 0$.

A fourth order patch has nine interior points, so there should be enough freedoms to match cubic patches on all sides. How this works out in practice is not known to the author, but a scheme in which alternate patches are of higher order is implemented in some commercially available systems.

The obvious advantage of this third method is that continuity is achieved between patches without going either to ninth order as in the first method or to rational functions as in the second method. How well it can be made to work for patches on an irregular layout of boundaries is not certain.

4. CORNER CROSS-DERIVATIVES

4.1 Physical significance

The significance of the interior control points $\underline{i}$ in a patch is not immediately clear. It is simpler to think in terms of the cross-derivative $\underline{p}_{uv}$ in the corner (0,0)

$$\begin{aligned}\underline{p}_{uv} &= \frac{\partial}{\partial u}(\underline{p}_v) = \frac{\partial}{\partial v}(\underline{p}_u)\\ &= 9(\underline{i} - \underline{a} - \underline{b} + \underline{p})\\ &= 3[3(\underline{i} - \underline{a}) - 3(\underline{b} - \underline{p})]\\ &= 3[3(\underline{i} - \underline{b}) - 3(\underline{a} - \underline{p})] \qquad (4.1)\end{aligned}$$

The components of $\underline{p}_{uv}$ in the plane of the surface at a corner $\underline{p}$ represent the rates of growth and turning of the tangent vectors $\underline{p}_u$ and $\underline{p}_v$ as one moves away from the corner, Figure 5. The component in the direction of the normal to the surface is related to the local twist of the surface, Figure 6.

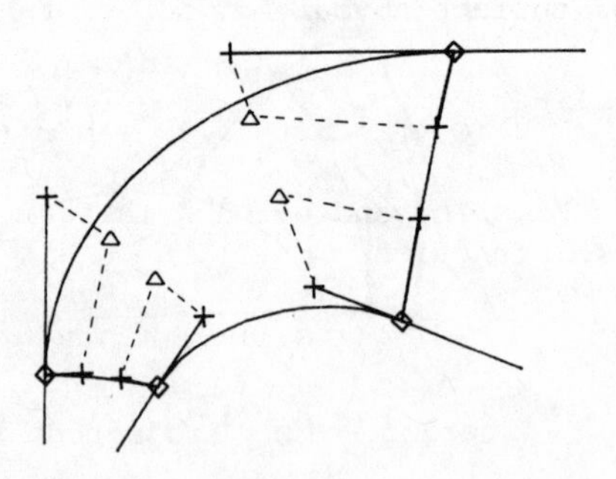

Fig. 5 Growth and Turning of Tangent Vectors

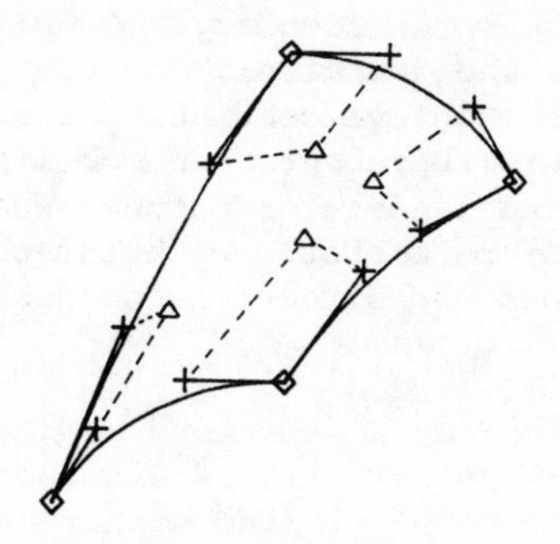

Fig. 6 Twist of Tangent Vectors

4.2 Component in the surface

The component of cross-derivative in the surface at the corners may be assigned by projecting the tangents to the patch boundaries at the adjacent corners onto the surface, Figure 7. The interior points i are placed on straight lines through the points of intersection of these tangents d or e and the associated control points a or b, as seen in this projection.

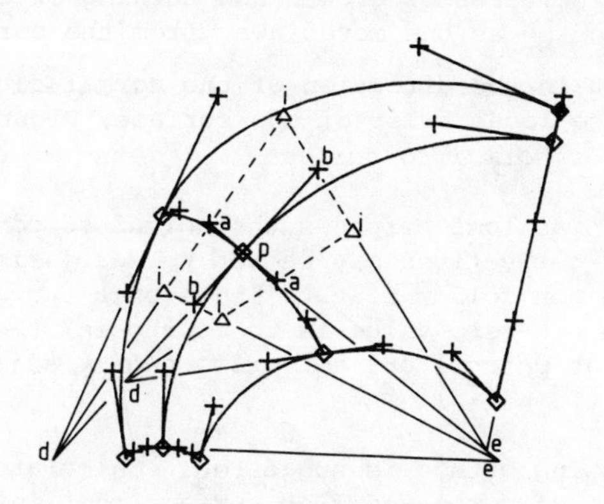

Fig. 7 Interior Points Projected onto Surface

This method of locating the interior points satisfies the requirements from Section 1 (b) for fairness, (c) for independence of patches and (d) to model surfaces of revolution correctly. It also meets the requirement of Sections 2 and 3.1 that the lines i - a - i and i - b - i are straight.

When describing the method of redistribution of parameter in 3.2 no mention was made of the interior points. As given the method works only when the ratio g is the same for the interior points as for the patch boundary. Consequently one of the sets of lines must be made parallel, and the other to converge on a point in order to achieve continuity across the boundaries, as in Figure 8. This represents a compromise on the interior smoothness and independece of patches when using the method of redistribution of parameter. With independent blending of tangent magnitudes and directions no such compromise is necessary.

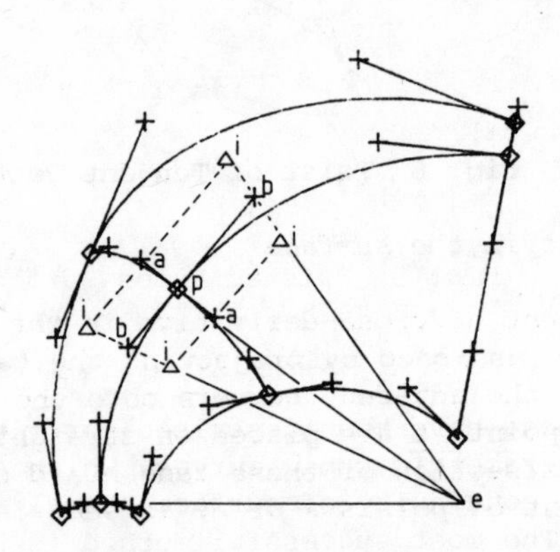

Fig. 8 Compromise on Positions of Interior Points

4.3 Component normal to surface

For many practical purposes the normal component of the corner cross-derivatives may be set to zero, since patch boundaries often follow feature lines on a surface and do not twist. If a non-zero value is to be chosen, then it must be such that flat patches are not distorted by adjacent twisted patches.

The following method is suggested. Calculate the constant rate of twist of a twisted flat surface containing adjacent corner points $\underline{p}_{00}$ and $\underline{p}_{01}$ and the control points $\underline{b}_{00}$ and $\underline{b}_{01}$ as in Figure 9. Repeat for each of the four points adjacent to the point under consideration. Set the corner twist to the lowest of the four rates of twist when they are of the same sign, or to zero when any are of opposite sign. Determine from the corner twist by how much the points $\underline{i}$ should be moved normal to the plane of the surface in the corner, in order to make the four interior points lie on the same twisted ruled surface.

As a result of this method, the interior shapes of patches are slightly dependent on adjacent patches, to the extent that a flattish patch may influence a more twisted adjacent patch.

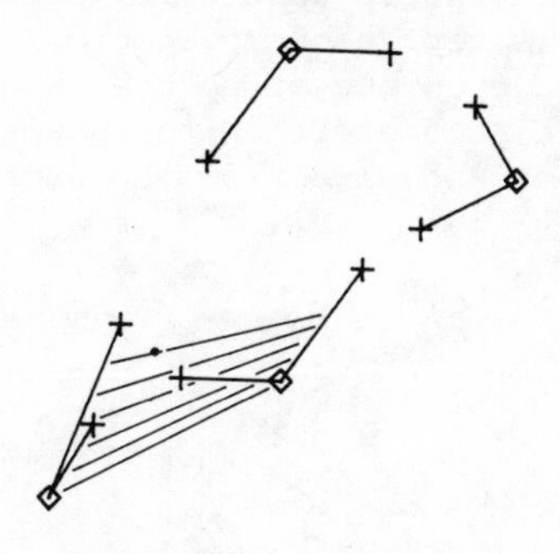

Fig. 9 Twisted Rules Surface

5. CONCLUSION

Three methods have been described for fitting patches to networks of parametric cubic curves, which combine internal fairness with continuity of surface normal across patch boundaries. The most successful method is to blend tangent magnitudes and directions independently, but points on the resulting ninth order patches are complicated to compute. A method which results in simpler calculations is to redistribute the parameter on one set of boundaries while blending tangent magnitudes and directions independently on the other. This combined method however requires a compromise on the positioning of the interior control points, and a possible consequent distortion of patches representing twisted ruled surfaces by adjacent patches.

REFERENCES

[1] Faux, I.D. and Pratt, M.J., *Computational Geometry for Design Manufacture*, Ellis Horwood Ltd., Chichester, 1979.

[2] Forrest, A.R., On Coons and Other Methods for the Representation of Curved Surfaces, Computer Graphics and Image Processing, Vol. 1, pp 341-359, 1972.

[3] Sturge, D.P. and Nicholls, M.S., Developments in the DUCT System of Computer Aided Engineering. In *Proc. 23rd International Machine Tool Design and Research Conf.* B.J. Davies (ed.), pp 257-261, UMIST/Macmillan Press, London, 1982.

[4] Welbourn, D.P., Computer Aided Engineering in the Foundary Industry, Giesserei, Düsseldorf, Vol. 69, No. 25 , pp. 734-744, 1982 (in German and English)

A CIRCLE DIAGRAM FOR LOCAL DIFFERENTIAL GEOMETRY

A.W. Nutbourne
(Cambridge University Engineering Department)

1. INTRODUCTION

The surface torsion of a surface curve is related to the curve torsion of the same curve by the formula

$$t = \tau + \phi' \tag{1.1}$$

where t is the surface torsion
τ is the curve torsion
ϕ is the angle between the curve normal $\underline{n}$ and the surface normal $\underline{N}$
$\phi' = \dfrac{d\phi}{ds}$, where s is arc length.

In words this equation can be stated as

The torsion of the surface = the torsion of the surface curve in its own right as a space curve + the relative torsion of the surface with respect to the curve.

The surface torsion (t) is more often called the geodesic torsion (τ_g) of the curve. It is convenient to suppress the word geodesic (probably introduced by Bonnet, 150 years ago) as it is much too specific in this context; and to replace it with the simpler idea of surface torsion. The symbol t need not cause confusion with a tangent vector $\underline{t}$ as the latter has unit magnitude and does not need the symbol t for its magnitude.

Because the space curve frame $[f] = [\underline{t}, \underline{n}, \underline{b}]$ and the surface curve frame $[F] = [\underline{t}, \underline{T}, \underline{N}]$ share the tangent vector $\underline{t}$; all the other four vectors lie in the normal plane of the curve and the two frames are related by a rotation ϕ about $\underline{t}$. It is easy to deduce that

$$\underline{N}' = \frac{d\underline{N}}{ds} = - n\underline{t} - t\underline{T} \tag{1.2}$$

where n is the surface normal curvature
t is the surface torsion
$\underline{t}$ is tangent vector to the surface in a specified direction
$\underline{T}$ is orthogonal to $\underline{t}$ in the surface tangent plane

n is more usually written as K_n, but it is helpful to have (1.2) stripped of suffices. In a similar fashion I use g to denote K_g, the geodesic curvature of a surface curve.

We may regard (1.2) as composed of two different physical components:

(1) $n(-\underline{t})$ expresses the pitching of the surface

(2) $t(-\underline{T})$ expresses the roll of the surface

This nautical analogue is readily understood if we regard $\underline{N}$ as a unit length mast on a ship sailing on the surface in a direction $\underline{t}$. When it has moved a distance ds from a point P, $d\underline{N}$ expresses the vector movement of the masthead with respect to its translated (but not rotated) position.

A principal direction may be defined as a direction in which the surface torsion is zero, and it is amply proved in elementary texts on differential geometry that there are two such directions (or four if you count a reversed direction as distinct) and that these are mutually orthogonal. This fundamental result is neither obvious, nor trivial to prove so the novice should not accept it without further anatomical dissection. The exceptional case of a spherical point, for which all directions are principal will be mentioned later.

Note that (1.2) for $t = 0$ yields Rodrigues' relation for a principal direction.

$$\underline{N}' = -n\underline{t} \tag{1.3}$$

Let the normal curvatures in the two principal directions be n_1 and n_2, choosing $n_1 \geq n_2$ to define which of the two directions shall be called $\underline{t}_1$, with $\underline{t}_2$ then defined by $\underline{t}_2 = \underline{T}_1$.

Let a general direction $\underline{t}$ in the surface tangent plane be inclined at an angle θ to $\underline{t}_1$. Then Euler's relation gives

$$n = n_1 \cos^2\theta + n_2 \sin^2\theta \tag{1.4}$$

and Sophie Germain's relation is

$$t = (n_2 - n_1) \sin\theta \cos\theta \tag{1.5}$$

The circle diagram welds these relations into a coherent whole.

2. THE CIRCLE DIAGRAM

Equations (1.4) and (1.5) can be restated in the form

$$n = \frac{1}{2}(n_1 + n_2) + \frac{1}{2}(n_1 - n_2)\cos 2\theta \qquad (2.1)$$

$$t = -\frac{1}{2}(n_1 - n_2)\sin 2\theta \qquad (2.2)$$

Eliminating θ gives

$$[n - \frac{1}{2}(n_1 + n_2)]^2 + t^2 = [\frac{1}{2}(n_1 - n_2)]^2, \qquad (2.3)$$

so that a plot of t against n is a circle centred on the mean normal curvature and of radius $\frac{1}{2}(n_1 - n_2)$. More simply, the plot is a circle centred on the n axis that passes through n_1 and n_2, as illustrated in Figure 1.

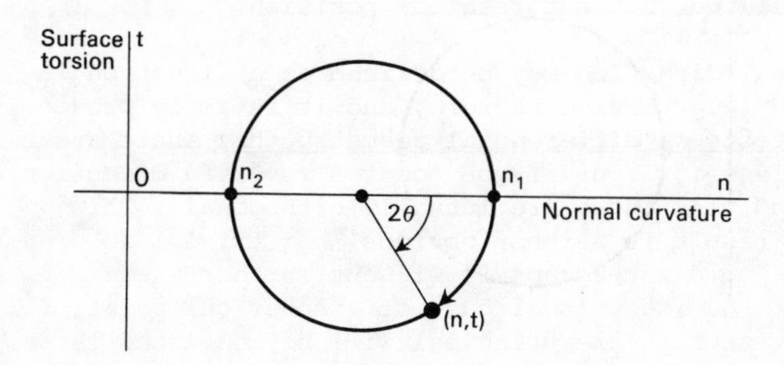

Fig. 1. The Circle Diagram

As θ increases (anticlockwise) the circle is traced clockwise through an angle 2θ, as may be observed from (2.1) and (2.2).

The Gaussian Curvature $K = n_1 n_2$ is a useful parameter to describe the various circle diagrams that are possible.

2.1 K Positive. Elliptic Point

n_1 and n_2 are either both positive (Fig. 1) or both negative (Fig. 2). The circle diagram will appear wholly in the right-hand half plane or the left-hand half plane and does not intersect the t axis.

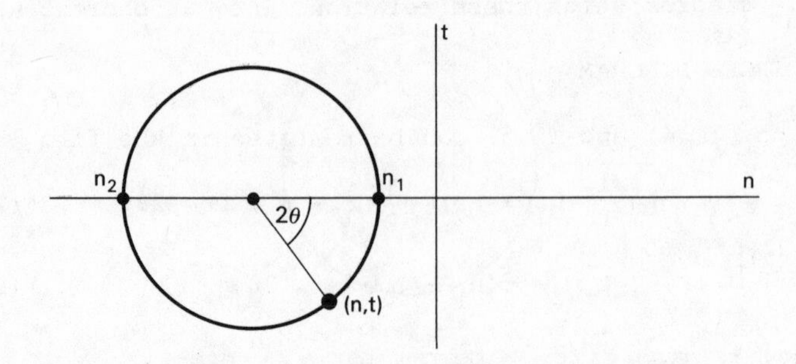

Fig. 2. The Alternative Circle Diagram for an Elliptic Point.

It is instructive to note that the length of the tangents to the circle from the origin O is $\sqrt{K}$. This follows immediately from the elementary theorem in geometry that the square of the tangent to a circle from a point P equals PA PB where PAB is a straight line intersecting the circle at A and B.

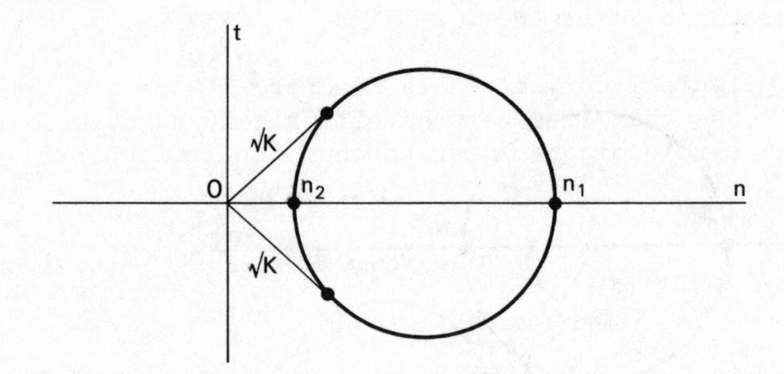

Fig. 3. The Gaussian Curvature

2.2 K negative. Hyperbolic Point

n_1 and n_2 are of opposite sign and because n_1 and n_2 were defined with the deliberate selection of principal directions so that $n_1 \geqslant n_2$, the circle diagram has one manifestation, see Fig. 4. Although there are now no tangents from O to the circle, the intercepts on the t axis take over this role and each has the length $\sqrt{-K}$. Note that $-K$ is a positive quantity so that $\sqrt{-K}$ is real.

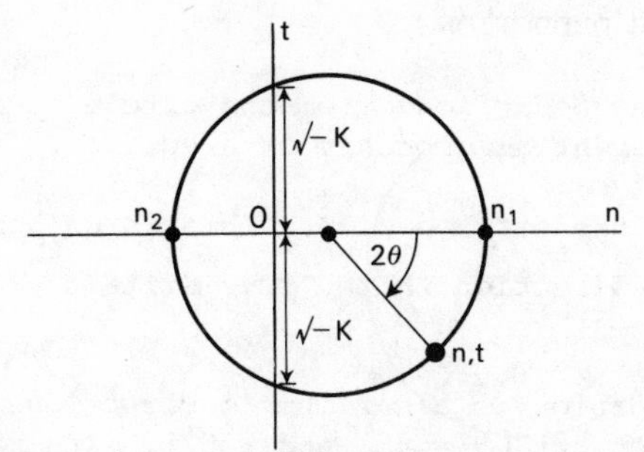

Fig. 4. The Circle Diagram for a Hyperbolic Point

It is now possible for n to be zero in two directions. These are called the asymptotic directions. Enneper's theorem that the geodesic torsion of an asymptotic curve is $\sqrt{-K}$ is now self-evident. It is also easy to see that the principal directions bisect the angles between the asymptotic directions, but not vice-versa.

2.3 K = O. Parabolic Point

One of the principal curvatures is now zero. Such a point is the special case that lies between elliptical and hyperbolic behaviour. The circle diagram just touches the torsion axis, see Figure 5.

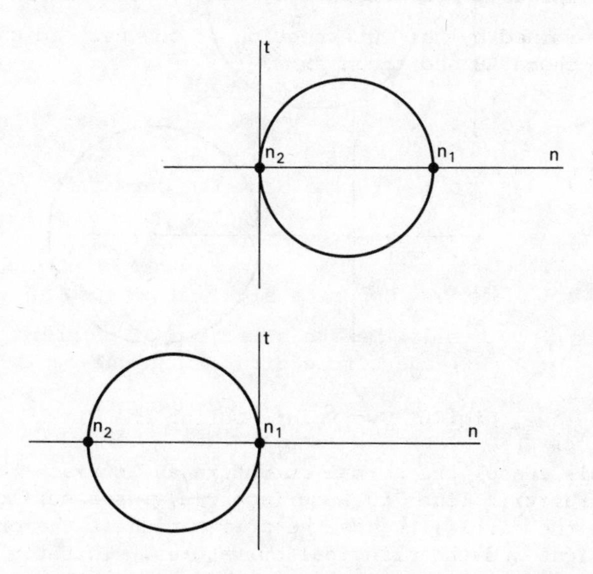

Fig. 5. The Circle Diagram for a Parabolic Point

There is only one asymptotic direction and it is coincident with a principal direction.

3. SOME SIMPLE DEDUCTIONS

The following deductions from the circle diagram illuminate the local differential geometry of a surface.

(1) n_1 and n_2 are extrema of n. This is a 'bonus' property of a principal direction though some writers promote it to a definition.

(2) Surface torsion is a maximum in directions that bisect the principal directions [$\theta = 45^\circ$ and 135°]. Its maximum value is the radius of the circle, $\frac{1}{2}(n_1 - n_2)$. The normal curvature in both these directions is $\frac{1}{2}(n_1 + n_2)$.

(3) The mean curvature is the distance of the centre of the circle from the origin. Note that it is the mean of the normal curvatures of any two directions that are orthogonal, as these plot at opposite ends of a diameter on the circle diagram. All points on a minimal surface have zero mean curvature. On such a surface the asymptotic directions are orthogonal and bisect the principal directions.

(4) Because the circle always has its centre on the n axis, only two points (n_A, t_A) (n_B, t_B) on the circle are needed to fix it uniquely [except when $n_A = n_B$]. The centre of the circle is determined by the intersection of the perpendicular bisector of the chord AB and the n axis.

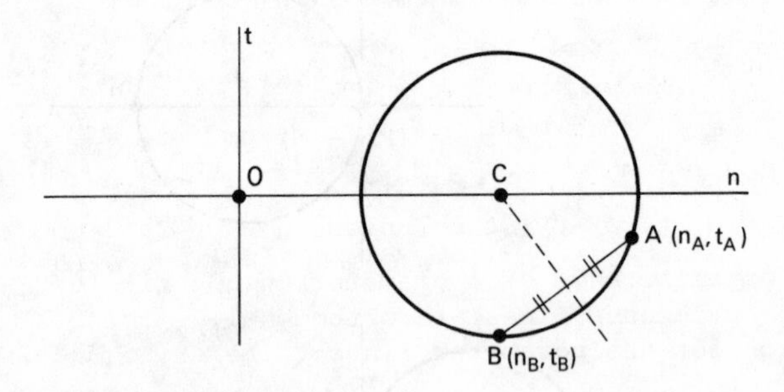

Fig. 6 Two Points Define the Circle

For this reason the normal curvature and surface torsion for the parametric lines on a surface $\underline{r}(u,v)$ are sufficient to define the circle, thence the orientation of the principal directions and the principal curvatures. This topic is dealt with more fully in Section 4. When $n_A = n_B$ the angle of intersection of the parameter lines is used to fix C.

(5) If $n_1 = n_2$, the circle shrinks to a point and all directions have zero surface torsion and are principal. The point is a spherical point.

(6) Two geodesic lines that emanate orthogonally from a point P have equal and opposite values of torsion τ. To see this note that $g = 0$ implies $\phi = 0$ giving $t = \tau$. Because the lines cut orthogonally their (n, t) points appear at opposite ends of a diameter of the circle, hence $t_1 = -t_2$. I omit discussion of the special case that occurs when $g = 0$ because $K = 0$ and ϕ is arbitrary.

4. ESTABLISHING THE CIRCLE DIAGRAM FROM THE PARAMETRIC LINES

If the surface is defined by the position vector $\underline{r}(u,v)$, then the parametric lines of constant v have tangent vectors in the direction $\underline{r}_u = \partial\underline{r}(u,v)/\partial u$; and the parametric lines of constant u have tangent vectors in the direction $\underline{r}_v = \partial\underline{r}(u,v)/\partial v$. Note that $\underline{r}_u$ and $\underline{r}_v$ are not of unit length nor are they necessarily orthogonal. It is shown in an Appendix that (n,t) for these parametric lines may be expressed in terms of the Gaussian parameters E, F, G; L, M, N, defined in the Appendix, as

$$(n_A;\ t_A) = [L/E;\ (EM - FL)/EH] \tag{4.1}$$

$$(n_B;\ t_B) = [N/G;\ (FN - GM)/GH] \tag{4.2}$$

Note that

$$t_A = (M - Fn_A)/H \tag{4.3}$$

$$t_B = (Fn_B - M)/H \tag{4.4}$$

If $F = 0$, $t_A = -t_B$; then unless $n_A = n_B$ the points A and B are at opposite ends of a diameter on the circle diagram so that the parameter lines are orthogonal. (This is a well-known condition for the parameter lines to be orthogonal.)

If additionally, $M = 0$ then t_A and t_B are zero so that the parameter lines are lines of curvature ($t = 0$ locus).

As an example we may take the doubly-ruled non-planar quadrilateral

$$\underline{r}(u,v) = \underline{a} + u\underline{p} + v\underline{q} + uv\,\underline{e} \tag{4.5}$$

illustrated in Figure 7.

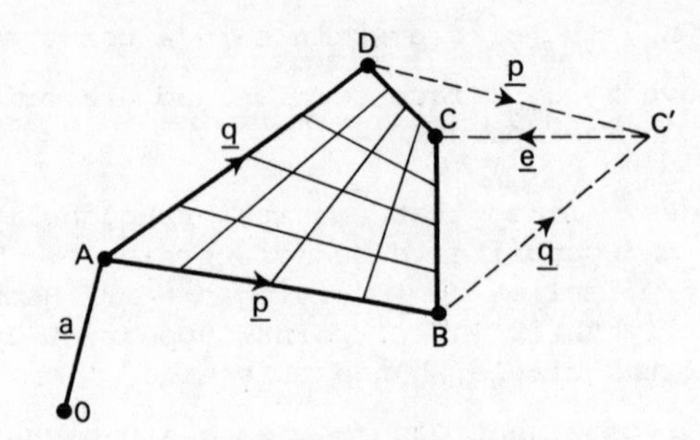

Fig. 7 Doubly-Ruled Quadrilateral

Then
$$\underline{r}_u = \underline{p} + v\underline{e}$$
$$\underline{r}_v = \underline{q} + u\underline{e}$$

Therefore
$$\underline{r}_u \times \underline{r}_v = \underline{p} \times \underline{q} + u\underline{p} \times \underline{e} + v\underline{e} \times \underline{q}$$
$$\underline{r}_{uu} = \underline{O}$$
$$\underline{r}_{uv} = \underline{e}$$
$$\underline{r}_{vv} = \underline{O}$$

$$E = \underline{r}_u.\underline{r}_u = p^2 + v^2e^2 + 2v\,\underline{p}.\underline{e}$$
$$F = \underline{r}_u.\underline{r}_v = \underline{p}.\underline{q} + uve^2 + u\underline{p}.\underline{e} + v\underline{q}.\underline{e}$$
$$G = \underline{r}_v.\underline{r}_v = q^2 + u^2e^2 + 2u\underline{q}.\underline{e}$$
$$L = -\underline{N}.\underline{r}_{uu} = O$$
$$N = -\underline{N}.\underline{r}_{vv} = O$$
$$M = -\underline{N}.\underline{r}_{uv} = -\frac{(\underline{r}_u \times \underline{r}_v).\underline{e}}{H} = -[\underline{p}, \underline{q}, \underline{e}]/H$$

where $H = |\underline{r}_u \times \underline{r}_v|$

Using (4.1) to (4.4)
$$(n_A, t_A) = (O, M/H)$$
$$(n_B, t_B) = (O, -M/H)$$

Since $n_A = n_B$ the circle diagram is not fully defined by these two points. If β is the angle between the parameter lines at a point, we have

$$\cot \beta = [\underline{r}_u \cdot \underline{r}_v]/|\underline{r}_u \times \underline{r}_v| = F/H$$

From Figure 8 we have OC = OB cot β so the centre of the circle is $[-MF/H^2, 0]$.

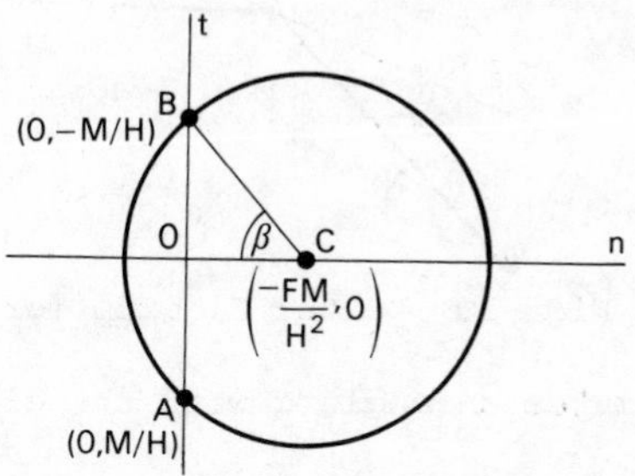

Fig. 8 Circle Diagram for Ruled Quadrilateral

The intercept on the t axis is $\sqrt{-K}$ giving $K = -M^2/H^2$ (This result is independent of the centre of the circle!)

The surface is wholly hyperbolic since K is always negative.

Note that the scalar triple product $[\underline{p}, \underline{q}, \underline{e}]$ is non-zero (except for a plane quadrilateral) so that $M \neq 0$. Even at points where F = 0 the parameter lines are not lines of curvature. The circle diagram shows this clearly. As $F \to 0$ the centre moves to O, but stays of finite radius.

5. THE VECTOR OQ ON THE CIRCLE DIAGRAM

Let Q be a point on the circle diagram, then we might reasonably ask what is the significance of vector OQ?

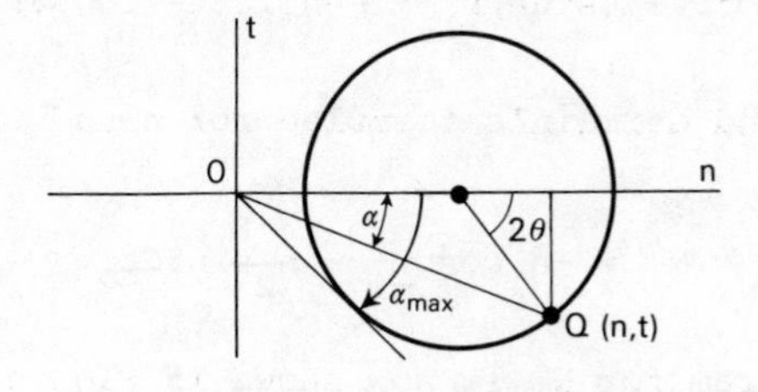

Fig. 9 The Vector OQ

Its length is $\sqrt{n^2 + t^2}$ and we may see from (1.2) that $|\underline{N}'| = \sqrt{n^2 + t^2}$. We may also correlate the angle α of Figure 9 with a vector diagram for $\underline{N}'$ given in Figure 10.

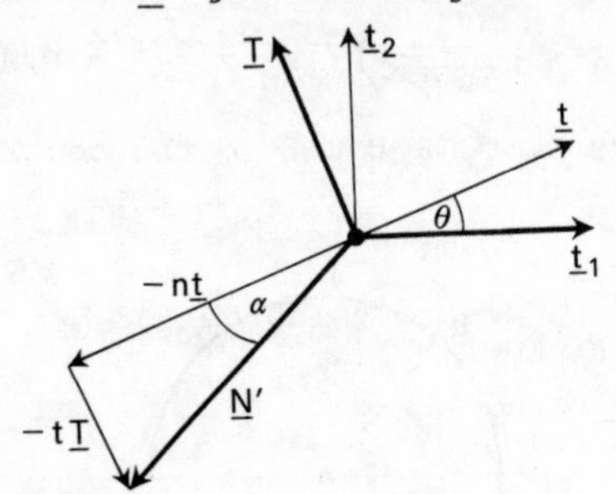

Fig. 10 Vector Diagram for $\underline{N}'$

We see that $\underline{N}'$ makes an angle α with the direction $-\underline{t}$.

The behaviour of $\underline{N}'$ at an elliptic point is now clear. The angle α is bounded between zero and the angle that the tangent to the circle diagram from O makes with the n axis. See α_{max} on figure.

$$\alpha_{max} = \sin^{-1}[\,|n_1 - n_2|/|n_1 + n_2|\,] \tag{5.1}$$

At a hyperbolic point α is unbounded and $\underline{N}'$ can escape more radically from the $\underline{t}$ direction.

6. A MORE USEFUL VECTOR DIAGRAM FOR $\underline{N}'$

The behaviour of $\underline{N}'$ is most easily appreciated if we replace $\underline{t}$ and $\underline{T}$ in (1.2)

$$\underline{t} = \underline{t}_1 \cos\theta + \underline{t}_2 \sin\theta \tag{6.1}$$

$$\underline{T} = \underline{t}_2 \cos\theta - \underline{t}_1 \sin\theta \tag{6.2}$$

Then $\underline{N}' = [-n\cos\theta + t\sin\theta]\underline{t}_1 + [-n\sin\theta - t\cos\theta]\underline{t}_2$

Using Euler's and Germain's formulae for n and t this reduces dramatically to

$$\underline{N}' = -n_1\cos\theta\,\underline{t}_1 - n_2\sin\theta\,\underline{t}_2 \tag{6.3}$$

The vector diagram for $\underline{N}'$ is now shown in Figure 11.

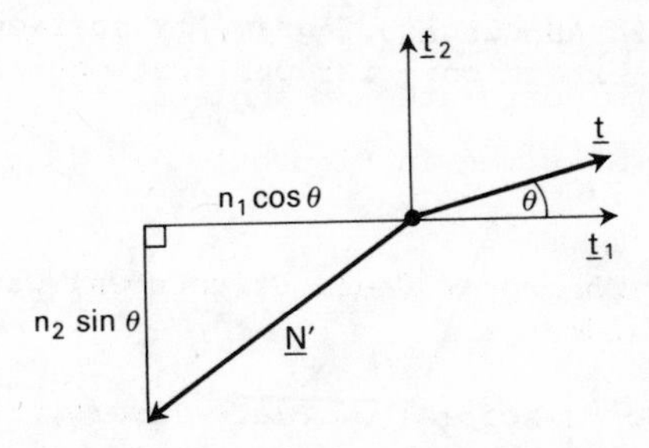

Fig. 11 Vector Diagram for $\underline{N}'$

It is now very clear why $\underline{N}'$ is only aligned with $\underline{t}$ for $\theta = 0^\circ$ or $\theta = 90^\circ$ unless $n_1 = n_2$.

7. CONCLUDING SECTION

Local differential geometry has two levels of sophistication: a somewhat crude level in which the true surface and its tangent plane are scarcely distinguished (a first order Taylor expansion?) and a more refined level in which all sorts of murky results about lines of curvature at umbilic points are revealed (higher order Taylor expansions?)

The circle diagram belongs to the first level. It provides a broad insight into the behaviour of a smooth surface but is not to be regarded as the whole truth.

It rams home the fact that surface torsion is a slave to the normal curvatures - caused by them - and is not an independent quantity.

No new results emerge: the exercise is just to gain insight and to help the newcomer to grasp the different behaviour for Gaussian Curvature negative, zero and positive.

I have not seen the vector diagram for the Normal Vector Derivative $\underline{N}'$ in any book. (See Figure 11) It displays $\underline{N}' = (-n_1\cos\theta)\underline{t}_1 - (n_2\sin\theta)\underline{t}_2$ instead of the more usual $\underline{N}' = -n\underline{t} - t\underline{T}$. It is an essential adjunct to the circle diagram which does not itself reveal the direction of $\underline{N}'$.

ACKNOWLEDGEMENT

Although I developed this teaching material independently, I have since discovered that Calladine [1] contains similar material in Chapter 5. Dr. Calladine tells me that his description of the Mohr's circle diagram for surface curvatures is dependent to some extent on prior publications of Dr. P.G. Lowe [2,3].

REFERENCES

[1] Calladine, C.R., *Theory of Shell Structures*, Cambridge University Press, 1983

[2] Lowe, P.G., *Basic Principles of Plate Theory*, Surrey University, 1982

[3] Lowe, P.G., A note of surface geometry with special reference reference to twist, Math. Proc. Camb. Phil. Soc. Vol. 87, pp. 481-487, 1980.

APPENDIX

$$E = \underline{r}_u \cdot \underline{r}_u$$

$$F = \underline{r}_u \cdot \underline{r}_v$$

$$G = \underline{r}_v \cdot \underline{r}_v$$

$$L = -\underline{N} \cdot \underline{r}_{uu}$$

$$M = -\underline{N} \cdot \underline{r}_{uv} = -\underline{N} \cdot \underline{r}_{vu}$$

$$N = -\underline{N} \cdot \underline{r}_{vv}$$

where

$$\underline{N} = [\underline{r}_u \times \underline{r}_v]/[\underline{r}_u \times \underline{r}_v]$$

These are the standard Gaussian parameters.

The surface normal curvature and surface torsion in a direction $h = dv/du$ is given by the formula

$$n = \frac{L + 2mh + Nh^2}{E + 2Fh + Gh^2} \qquad (A.1)$$

$$t = \frac{(EM - FL) + (EN - GL)h + (FN - GM)h^2}{H(E + 2Fh + Gh^2)} \qquad (A.2)$$

$h = 0$ gives $(n_A; t_A)$ for parameter lines v constant

$h = \infty$ gives $(n_B; t_B)$ for parameter lines u constant.

The numerator of (A.2) is written in determinant form very neatly

$$\begin{vmatrix} 1 & -h & h^2 \\ E & F & G \\ L & M & N \end{vmatrix} \tag{A.3}$$

If E: F: G = L: M: N the surface torsion is zero in all directions and the circle diagram must shrink to a point. This is the condition for a spherical point on the surface.

ELEMENTARY EXPOSITION OF DIFFERENTIAL FORMS

L.M. Woodward
(University of Durham)

1. INTRODUCTION

Differential forms were introduced by Élie Cartan in his paper "Sur certaines expressions différentielles et le probleme de Pfaff" published in 1899. (See [1]). The problem with which he was concerned is, in its simplest form, that of deciding when a differential expression of the form $\alpha = Pdx + Qdy + Rdz$ can be written in the form $\alpha = fdg$ for suitable functions f and g. In tackling this problem Cartan developed a calculus of such expressions called the exterior differential calculus which combines some simple algebraic operations with an operation of differentiation called exterior differentiation. This calculus has turned out to be a particularly powerful tool in several areas of mathematics but perhaps its most spectacular successes have been in differential geometry, where Cartan used differential forms in conjunction with moving frames to achieve a simple and elegant treatment of the equations of structure for differentiable manifolds. The strength of this approach is particularly well seen in Cartan's celebrated work on Lie groups.

In a first encounter with differential forms it is perhaps easiest to consider them as integrands of multiple integrals. Thus we have in 3 dimensions:

1-forms: integrands $fdx + gdy + hdz$ of line integrals;

2-forms: integrands $fdydz + gdzdx + hdxdy$ of surface integrals;

3-forms: integrands $fdxdydz$ of volume integrals.

The integrals concerned involve the consideration of orientation. For example in the case of a surface integral $\int_S fdydz + gdzdx + hdxdy$ a change in the orientation of the surface results in a change of sign of the integral. In the exterior differential calculus the orientation is incorporated through a certain anti-symmetry in the algebra. Namely we write $dxdy = -dydx$, $dxdydz = -dydxdz = dydzdx = -dzdydx = dzdxdy = -dxdzdy$. (This forces us to write $dxdx = 0$, $dxdxdy = 0$ and so on but it should be noted that such expressions never appear in integrands). If we consider the case of the plane with cartesian coordinates x,y then there are two choices of orientation: the standard "anticlockwise" orientation, and the opposite "clockwise" orientation. We think of these as represented by dxdy and dydx respectively.

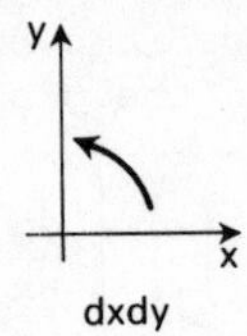

Fig. 1 Anticlockwise orientation

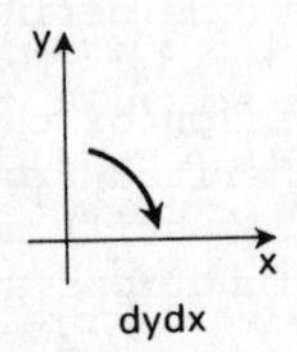

Fig. 2 Clockwise orientation

We have a similar situation in 3 dimensions with "right-handed" and "left-handed" orientations.

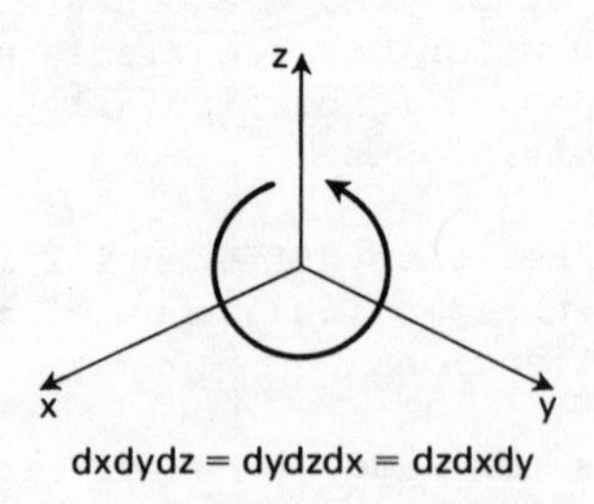

Fig. 3 Right-handed orientation

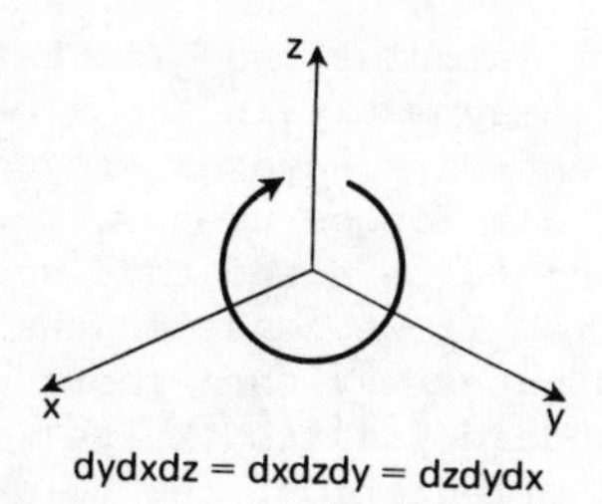

Fig. 4 Left-handed orientation

To see how well this algebraic formalism handles orientation consider Green's Theorem in the plane.

$$\int_C Pdx + Qdy = \iint_D \left(\frac{\partial Q}{\partial x} - \frac{\partial P}{\partial y} \right) dx\, dy \tag{1.1}$$

Here P,Q are smooth functions defined on a domain D in the plane bounded by a curve C. We give D the orientation inherited from the standard orientation of the plane and C the anti-clockwise orientation inherited from D. If these orientations are both changed then (1.1) still holds. Now let us relabel the coordinates and the functions as follows, as shown in Figure 5.

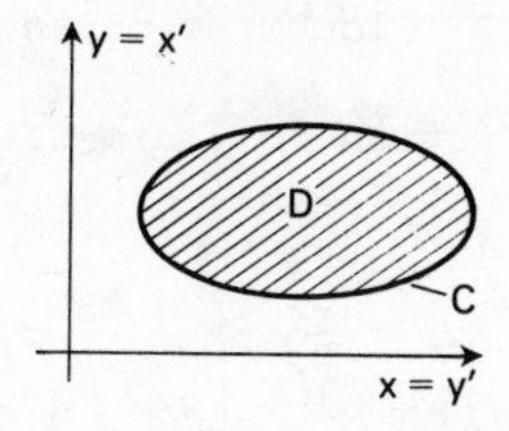

Fig. 5

$$x' = y,\ y' = x \tag{1.2}$$

$$P'(x', y') = Q(x, y),\ Q'(x', y') = P(x, y) \tag{1.3}$$

Thus we have

$$\int_C Pdx + Qdy = \int_C P'dx' + Q'dy' , \tag{1.4}$$

while

$$\iint_D \left(\frac{\partial Q}{\partial x} - \frac{\partial P}{\partial y} \right) dx\, dy = \iint_D \left(\frac{\partial P'}{\partial y'} - \frac{\partial Q'}{\partial x'} \right) dy'dx' \qquad (1.5)$$

Thus from (1.4) and (1.5) we have

$$\int_C P'dx' + Q'dy' = -\iint_D \left(\frac{\partial Q'}{\partial x'} - \frac{\partial P'}{\partial y'} \right) dy'dx'. \qquad (1.6)$$

If we do not invoke the anti-symmetry we have to remark that the change of variable has caused a change of orientation which results in the minus sign on the right-hand side of (1.6). If on the other hand we write $dx'dy' = - dy'dx'$ then (1.6) becomes

$$\int_C P'dx' + Q'dy' = \iint_D \left(\frac{\partial Q'}{\partial x'} - \frac{\partial P'}{\partial y'} \right) dx'dy' \qquad (1.7)$$

and the change of orientation is taken care of by the algebraic formalism.

To remind us of this anti-symmetry we write $dx \wedge dy$ (pronounced dx wedge dy) in place of dxdy and $dx \wedge dy \wedge dz$ in place of dxdydz.

The next important ingredient is that of the exterior derivative. Notice that in Green's Theorem above the 1-form $\alpha = Pdx + Qdy$ gives rise to a 2-form $d\alpha = \left(\frac{\partial Q}{\partial x} - \frac{\partial P}{\partial y} \right) dx \wedge dy$.

Now

$$dP = \frac{\partial P}{\partial x} dx + \frac{\partial P}{\partial y} dy \qquad (1.8)$$

$$dQ = \frac{\partial Q}{\partial x} dx + \frac{\partial Q}{\partial y} dy \qquad (1.9)$$

so that

$$dP \wedge dx + dQ \wedge dy = \left(\frac{\partial P}{\partial x} dx + \frac{\partial P}{\partial y} dy \right) \wedge dx + \left(\frac{\partial Q}{\partial x} dx + \frac{\partial Q}{\partial y} dy \right) \wedge dy$$

$$= \left(\frac{\partial Q}{\partial x} - \frac{\partial P}{\partial y} \right) dx \wedge dy.$$

Thus if $\alpha = Pdx + Qdy$ then $d\alpha = dP\wedge dx + dQ\wedge dy$ and we call $d\alpha$ the exterior derivative of α. Similarly in 3-dimensions if α is the 1-form $fdx + gdy + hdz$ then we define $d\alpha$ to be the 2-form $df\wedge dx + dg\wedge dy + dh\wedge dz$ so that

$$d\alpha = \left(\frac{\partial h}{\partial y} - \frac{\partial g}{\partial z}\right)dy\wedge dz + \left(\frac{\partial f}{\partial z} - \frac{\partial h}{\partial x}\right)dz\wedge dx + \left(\frac{\partial g}{\partial x} - \frac{\partial f}{\partial y}\right)dx\wedge dy;$$

while if $\alpha = fdy\wedge dz + gdz\wedge dx + hdx\wedge dy$ is a 2-form then we define $d\alpha = df\wedge dy\wedge dz + dg\wedge dz\wedge dx + dh\wedge dx\wedge dy$ so that

$$d\alpha = \left(\frac{\partial f}{\partial x} + \frac{\partial g}{\partial y} + \frac{\partial h}{\partial z}\right) dx\wedge dy\wedge dz.$$

From this and the fact that $df = \frac{\partial f}{\partial x}dx + \frac{\partial f}{\partial y}dy + \frac{\partial f}{\partial z}dz$ it is clear that there is a close relationship between the exterior differential operator d and the operators grad, div and curl. Furthermore Green's Theorem in the plane can be written very simply as

$$\int_{\partial D} \alpha = \int_D d\alpha , \tag{1.10}$$

where $\partial D = C$ is the boundary of the region D. The Divergence Theorem and the classical Stokes Theorem can also be written in the form (1.10); indeed they are special cases of a single generalized Stokes Theorem. These examples are typical of the way in which familiar expressions and results from the calculus of several variables can be written in a much more elegant and simplified way using differential forms. Another prime example is that of the Frobenius Integrability Theorem for Pfaffian systems, (see [3]).

This essay is intended only as a very brief and elementary introduction to differential forms. The interested reader will find more details and a wide variety of examples in the references, especially [3], which also contains an extensive bibliography.

Notation

We use $\mathbb{R}^n$ to denote real n-space whose elements are n-tuples of real numbers and we shall be mainly concerned with $\mathbb{R}^1$, $\mathbb{R}^2$ and $\mathbb{R}^3$, the real line, the plane and 3-dimensional space respectively. To simplify the notation we shall write $x^1, .., x^n$ for the coordinate functions corresponding to the basis vectors $(1,0,..,0),\ldots,(0,\ldots,0,1)$, so that for

$\mathbb{R}^3$ for example, we shall use x^1, x^2, x^3 in place of x, y, z.

Also we shall use the summation convention in which repeated indices are summed over so that, for example, $a_i x^i$ stands for $a_1 x^1 + .. + a_n x^n$, and $a_{ij} x^i x^j$ stands for $a_{11}(x^1)^2 + a_{12} x^1 x^2 + \ldots + a_{nn-1} x^n x^{n-1} + a_{nn}(x^n)^2$.

2. FORMAL DEFINITION AND ALGEBRAIC PROPERTIES

Following Cartan we begin by giving a purely formal definition of differential forms. The reader who wishes to understand what the symbols actually represent should turn to §7. In all that follows all functions will be assumed to be C^∞ although for the most part it suffices to assume them to be C^r for small values of r.

A <u>0-form</u> on $\mathbb{R}^n$ is a function $f = f(x^1, \ldots, x^n)$. A <u>1-form</u> on $\mathbb{R}^n$ is a differential expression $f_1 dx^1 + .. + f_n dx^n$, where $f_1, .., f_n$ are functions of $x^1, .., x^n$. Using the summation convention we write this as $f_i dx^i$.

A p-form on $\mathbb{R}^n$ is an expression of the form $f_{i_1 \ldots i_p} dx^{i_1} \wedge \ldots \wedge dx^{i_p}$.

We assume that the interchange of any two symbols in an expression such as $dx^{i_1} \wedge .. \wedge dx^{i_p}$ changes the sign so that, for example,

$$dx^i \wedge dx^j = - dx^j \wedge dx^i \qquad (2.1)$$

for all indices i, j. In particular if any two of the indices $i_1, \ldots, i_p$ are equal then $dx^{i_1} \wedge \ldots \wedge dx^{i_p} = 0$. Thus is α is a p-form on $\mathbb{R}^n$ with $p > n$ then $\alpha = 0$. Also if σ is a permutation of 1, .., n then

$$dx^{\sigma(1)} \wedge \ldots \wedge dx^{\sigma(n)} = (\text{sign } \sigma) . dx^1 \wedge \ldots \wedge dx^n \qquad (2.2)$$

where sign $\sigma = \pm 1$ according as σ is an even or odd permutation.

If $\alpha = f_{i_1..i_p} dx^{i_1}\wedge..\wedge dx^{i_p}$ is a p-form we call p the <u>degree</u> of α and write deg $\alpha = p$.

There are three algebraic operations defined on differential forms:

(i) addition: if $\alpha = f_{i_1...i_p} dx^{i_1}\wedge..\wedge dx^{i_p}$,

$$\beta = g_{i_1..i_p} dx^{i_1}\wedge...\wedge dx^{i_p}$$

then

$$\alpha + \beta = (f_{i_1..i_p} + g_{i_1..i_p})dx^{i_1}\wedge...\wedge dx^{i_p}.$$

(Note $\alpha + \beta$ is defined only for α and β of the same degree).

(ii) scalar multiplication (by functions):

if $\alpha = f_{i_1..i_p} dx^{i_1}\wedge..\wedge dx^{i_p}$ and λ is a function then

$$\lambda\alpha = \lambda f_{i_1..i_p} dx^{i_1}\wedge..\wedge dx^{i_p}.$$

(iii) exterior multiplication: if $\alpha = f_{i_1..i_p} dx^{i_1}\wedge...\wedge dx^{i_p}$ and $\beta = g_{j_1..j_q} dx^{j_1}\wedge..\wedge dx^{j_q}$ then

$$\alpha \wedge \beta = f_{i_1..i_p} g_{j_1..j_q} dx^{i_1}\wedge..\wedge dx^{i_p}\wedge dx^{j_1}\wedge..\wedge dx^{j_q}.$$

Observe that (ii) is a special case of (iii) where one of the forms is a 0-form. Also (i) and (ii) satisfy all the algebraic rules which hold for addition and scalar multiplication of vectors, while exterior multiplication satisfies:

(a) (associative law) $\alpha \wedge (\beta\wedge\gamma) = (\alpha\wedge\beta)\wedge\gamma$.

(b) (distributive law) $\alpha\wedge (\beta + \gamma) = \alpha\wedge\beta + \alpha\wedge\gamma$.

(c) $\alpha\wedge\beta = (-1)^{\deg\alpha.\deg\beta}\beta\wedge\alpha$.

(d) $\deg(\alpha\wedge\beta) = \deg\alpha + \deg \beta$.

In view of the above it follows that in $\mathbb{R}^3$ we can write each differential form uniquely as follows:

0-forms f,

1-forms $f_1dx^1 + f_2dx^2 + f_3dx^3$,

2-forms $f_1dx^2\wedge dx^3 + f_2dx^3\wedge dx^1 + f_3dx^1\wedge dx^2$,

3-forms $f\,dx^1\wedge dx^2\wedge dx^3$,

while all p-forms are zero for $p \geqslant 4$.

Examples

(i) If $\alpha = f_1dx^1 + f_2dx^2 + f_3dx^3$, $\beta = g_1dx^1 + g_2dx^2 + g_3dx^3$

then

$$\alpha + \beta = (f_1 + g_1)dx^1 + (f_2 + g_2)dx^2 + (f_3 + g_3)\,dx^3$$

while

$$\alpha\wedge\beta = (f_2g_3 - f_3g_2)\,dx^2\wedge dx^3 + (f_3g_1 - f_1g_3)dx^3\wedge dx^1$$
$$+ (f_1g_2 - f_2g_1)dx^1\wedge dx^2.$$

(ii) If $\alpha = f_1dx^2\wedge dx^3 + f_2dx^3\wedge dx^1 + f_1dx^1\wedge dx^2$,

$\beta = g_1dx^1 + g_2dx^2 + g_3dx^3$ then

$\alpha\wedge\beta = \beta\wedge\alpha = (f_1g_1 + f_2g_2 + f_3g_3)dx^1\wedge dx^2\wedge dx^3$.

(iii) If $\alpha = f_1dx^1 + f_2dx^2 + f_3dx^3$, $\beta = g_1dx^1 + g_2dx^2 + g_3dx^3$

$\gamma = h_1dx^1 + h_2dx^2 + h_3dx^3$ then

$$\alpha\wedge\beta\wedge\gamma = \det\begin{pmatrix} f_1 & g_1 & h_1 \\ f_2 & g_2 & h_2 \\ f_3 & g_3 & h_3 \end{pmatrix} dx^1\wedge dx^2\wedge dx^3.$$

3. EXTERIOR DIFFERENTIATION

If $\alpha = f_{i_1..i_p} dx^{i_1}\wedge..\wedge dx^{i_p}$ is a p-form we define its <u>exterior derivative</u> $d\alpha$ by

$$d\alpha = \frac{\partial f_{i_1..i_p}}{\partial x^i} dx^i \wedge dx^{i_1}\wedge..\wedge dx^{i_p}.$$

In particular if $\alpha = f$ is a 0-form then

$$d\alpha = \frac{\partial f}{\partial x^i} dx^i$$

is the differential df of f. The operator d is called the <u>exterior differential operator</u> and the operation of replacing a p-form α by its exterior derivative $d\alpha$ is called <u>exterior differentiation</u>.

<u>Example</u> If $\alpha = (x+y)dx + zdy + ydz$ then $d\alpha = - dx\wedge dy$.

The following properties of the operator d are easily verified:

(i) $\deg(d\alpha) = \deg\alpha + 1$,

(ii) if f is a 0-form then $df = \frac{\partial f}{\partial x^i} dx^i$

(iii) $d(\alpha + \beta) = d\alpha + d\beta$.

(iv) (Leibniz formula) $d(\alpha\wedge\beta) = d\alpha\wedge\beta + (-1)^{\deg\alpha}\alpha\wedge d\beta$.

(v) $d^2 = 0$; i.e. $d(d\alpha) = 0$.

To establish (v) observe that if $\alpha = f_{i_1..i_p} dx^{i_1}\wedge..\wedge dx^{i_p}$ then

$$d(d\alpha) = d\left\{\frac{\partial f_{i_1..i_p}}{\partial x^i} dx^i\wedge dx^{i_1}\wedge..\wedge dx^{i_p}\right\}$$

$$= \frac{\partial^2 f_{i_1..i_p}}{\partial x^j \partial x^i} dx^j\wedge dx^i\wedge dx^{i_1}\wedge..\wedge dx^{i_p}.$$

But

$$\frac{\partial^2 f}{\partial x^j \partial x^i} dx^j \wedge dx^i = \frac{\partial^2 f}{\partial x^i \partial x^j} dx^j \wedge dx^i$$

$$= - \frac{\partial^2 f}{\partial x^i \partial x^j} dx^i \wedge dx^j \text{ by (2.1)}$$

so $\frac{\partial^2 f}{\partial x^j \partial x^i} dx^j \wedge dx^i = 0$. Thus $d(d\alpha) = 0$.

The exterior derivative is characterized by properties (i) - (v) in the sense that if $\tilde{d}$ is an operator satisfying those properties then $\tilde{d} = d$.

Property (v) has a converse known as the Poincaré Lemma.

Poincaré Lemma If α is a p-form on $\mathbb{R}^n$ such that $d\alpha = 0$ then $\alpha = d\beta$ for some (p-1)-form β.

Remark. This lemma is false in general if $\mathbb{R}^n$ is replaced by some subspace of $\mathbb{R}^n$. For example consider the 1-form $= \frac{xdy - ydx}{x^2 + y^2}$ defined on $\mathbb{R}^2 \setminus \{(0,0)\}$. This satisfies $d\alpha = 0$ but there is no 0-form f defined on $\mathbb{R}^2 \setminus \{(0,0)\}$ such that $\alpha = df$. For otherwise, applying the Fundamental Theorem of Calculus to the integral of df around the unit circle S^1, we would have $\int_{S^1} \alpha = 0$, whereas in terms of polar coordinates we have $\alpha = d\theta$ so that $\int_{S^1} \alpha = 2\pi$. (Here θ is not differentiable over the whole of $\mathbb{R}^2 \setminus \{(0,0)\}$, of course). The failure of the Poincaré Lemma to hold in general can be used to obtain topological information - in the above example it shows that $\mathbb{R}^2 \setminus \{(0,0)\}$ has a "hole". This has been refined into a very powerful tool called de Rham cohomology theory which uses differential forms to investigate topological properties. See [3,5] for further details.

Change of variables

Suppose now that $x^1, .., x^n$ are functions of m real variables $y^1, .., y^m$. Then a p-form $f_{i_1 .. i_p} dx^{i_1} \wedge .. \wedge dx^{i_p}$ in the variables $x^1,..,x^n$ determines a p-form in the variables $y^1,..., y^m$ according to the following rules: we replace $f_{i_1..i_p}(x^1,..,x^n)$ by

$$g_{i_1..i_p}(y^1,...,y^m) = f_{i_1..i_p}(x^1(y^1,..,y^m), ...,x^n(y^1,..., y^m))$$

and replace dx^i by $\frac{\partial x^i}{\partial y^j} dy^j$. Thus $f_{i_1..i_p} dx^{i_1}\wedge..\wedge dx^{i_p}$ becomes

$$g_{i_1..i_p} \frac{\partial x^{i_1}}{\partial y^{j_1}} \cdots \frac{\partial x^{i_p}}{\partial y^{j_p}} dy^{j_1}\wedge..\wedge dy^{j_p}.$$

In particular when m = n we have

$$dx^1 \ldots dx^n = \frac{\partial x^1}{\partial y^{j_1}} .. \frac{\partial x^n}{\partial y^{j_1}} .. dy^{j_1}\wedge..\wedge dy^{j_n}$$

$$= \sum_\sigma \frac{\partial x^1}{\partial y^{\sigma(1)}} \cdots \frac{\partial x^n}{\partial y^{\sigma(n)}} dy^{\sigma(1)}\wedge..\wedge dy^{\sigma(n)}$$

where the summation is over all permutations σ of 1, .., n

$$= \sum_\sigma (\text{sign } \sigma) \frac{\partial x^1}{\partial y^{\sigma(1)}} \cdots \frac{\partial x^n}{\partial y^{\sigma(n)}} dy^1\wedge..\wedge dy^n$$

by (2.2)

$$= \det\left(\frac{\partial x^i}{\partial y^i}\right) dy^1\wedge..\wedge dy^n.$$

<u>Example</u> Let $x = r \cos \theta$, $y = r \sin \theta$. Then

$$dx = \cos \theta \, dr - r \sin \theta d\theta$$

$$dy = \sin \theta \, dr + r \cos \theta d\theta$$

so $dx\wedge dy = (\cos \theta \, dr - r \sin \theta d\theta)(\sin \theta \, dr + r \cos \theta d\theta)$

$$= r (\cos^2\theta + \sin^2 \theta) dr \, d\theta.$$

Thus $dx\wedge dy = r \, dr\wedge d\theta$.

4. THE HODGE STAR OPERATOR

We now define an operator $*$, depending on a choice of orientation and a metric on $\mathbb{R}^n$, which assign to a p-form α on $\mathbb{R}^n$ an (n-p)-form $*\alpha$ on $\mathbb{R}^n$ and which satisfies the two conditions

*(i) $*(\alpha+\beta) = *\alpha + *\beta$,

*(ii) $*(\lambda\alpha) = \lambda(*\alpha)$,

for all p-forms α, β and all real-valued functions λ. To simplify matters we will consider first the case of the standard metric $ds^2 = (dx^1)^2 + \ldots + (dx^n)^2$ and the orientation $x^1, \ldots, x^n$ of the coordinates. In this case we define

$$*dx^{i_1}\wedge\ldots\wedge dx^{i_p} = \varepsilon dx^{j_1}\wedge\ldots\wedge dx^{j_{n-p}},$$

where $(i_1, \ldots, i_p, j_1, \ldots, j_{n-p}) = (\sigma(1), \ldots, \sigma(n))$ for some permutation σ of $1, \ldots, n$ and $\varepsilon = \text{sign}\,\sigma$. Equivalently

$$*dx^{i_1}\wedge\ldots\wedge dx^{i_p} = \varepsilon dx^{j_1}\wedge\ldots\wedge dx^{j_{n-p}}$$

where $\varepsilon = \pm 1$ and

$$dx^{i_1}\wedge\ldots\wedge dx^{i_p}\wedge *(dx^{i_1}\wedge\ldots\wedge dx^{i_p}) = dx^1\wedge\ldots\wedge dx^n.$$

Thus, for example,

$$*dx^i = (-1)^{i-1} dx^1\wedge\ldots\wedge dx^{i-1}\wedge dx^{i+1}\wedge\ldots\wedge dx^n.$$

Also we define $*dx^1\wedge\ldots\wedge dx^n = 1$, the constant function which takes the value 1 at every point, and $*1 = dx^1\wedge\ldots\wedge dx^n$. The n-form $*1$ is called the <u>volume form</u>. If the opposite orientation (corresponding to the ordering $x_2, x_1, x_3, \ldots, x_n$) is used the corresponding $*$ operator is the negative of the one here and the volume form in this case is $-dx^1\wedge dx^2\wedge\ldots\wedge dx^n$.

<u>Examples</u> (i) $n = 2$. Here $*f = f\, dx^1\wedge dx^2$,

$$*(f_1 dx^1 + f_2 dx^2) = -f_2 dx^1 + f_1 dx^2, \quad *f\, dx^1\wedge dx^2 = f.$$

(ii) $n = 3$. Here $*f = f dx^1 \wedge dx^2 \wedge dx^3$,

$$*(f_1 dx^1 + f_2 dx^2 + f_3 dx^3) = f_1 dx^2 \wedge dx^3 + f_2 dx^3 \wedge dx^1 + f_3 dx^1 \wedge dx^2$$

$$*f dx^1 \wedge dx^2 \wedge dx^3 = f.$$

Frequently, especially in the context of differential geometry, one needs to consider more general metrics of the form $ds^2 = g_{ij} dx^i dx^j$ (where $g_{ij} = g_{ji}$). In such cases we write the metric in the form $ds^2 = (\theta^1)^2 + \ldots + (\theta^n)^2$ where $\theta^1, \ldots, \theta^n$ are 1-forms. (This is always possible but is particularly easy to do in the case where the matrix (g_{ij}) is diagonal). In this case we define

$$*\theta^{i_1} \wedge \ldots \wedge \theta^{i_p} = \varepsilon \theta^{j_1} \wedge \ldots \wedge \theta^{j_{n-p}},$$

where $(i_1, \ldots, i_p, j_1, \ldots, j_{n-p}) = (\sigma(1), \ldots, \sigma(n))$ for some permutation σ of $1, \ldots, n$ and $\varepsilon = \text{sign}\, \sigma$. In this case the volume form $*1 = \theta^1 \wedge \ldots \wedge \theta^n$. (It is not hard to check that if $ds^2 = g_{ij} dx^i dx^j$ then $*1 = \pm\sqrt{\det(g_{ij})}\, dx^1 \wedge \ldots \wedge dx^n$ and $\theta^1, \ldots, \theta^n$ are said to determine the same orientation as $x^1, \ldots, x^n$ if the sign is positive). Also it is easy to check that

*(iii) $**\alpha = (-1)^{p(n-p)} \alpha$, where $p = \deg \alpha$.

Example $ds^2 = dr^2 + r^2 d\theta^2$. Here we write $\theta^1 = dr, \theta^2 = r d\theta$ so that the volume form, (or area form), is given by $\theta^1 \wedge \theta^2 = r dr \wedge d\theta$. Then $*f = fr\, dr \wedge d\theta$, $*(f\, dr + g\, d\theta) = -\frac{g}{r}\, dr + rf\, d\theta$, $*f dr \wedge d\theta = \frac{f}{r}$.

Remark. The duality between p-forms and (n-p)-forms has a geometric counterpart in the form of the dual triangulation of a triangulation. The full topological importance of this duality can be seen in cohomology theory where it is known as Poincaré duality. See [5].

5. GRAD, DIV AND CURL

We now show how to express grad, div and curl in terms of the exterior derivative. Again, for simplicity, we will first describe the situation for the case where $\mathbb{R}^n$ has the standard metric $ds^2 = (dx^1)^2 + \ldots + (dx^n)^2$. Given the metric there is a 1-1 correspondence between vector fields and 1-forms which we write as

$$\underline{a} = (a^1, .., a^n) \longleftrightarrow \alpha = a_1 dx^1 + .. + a_n dx^n,$$

where $a_i = a^i$. (The need to use subscripts and superscripts is dictated by the conventions of tensor calculus. If the metric is $ds^2 = g_{ij} dx^i dx^j$ then $a_i = g_{ij} a^j$. Here we have $g_{ij} = \delta_{ij}$, the Kronecker δ.)

If f is a function then

$$\text{grad } f = \left(\frac{\partial f}{\partial x^1}, \ldots, \frac{\partial f}{\partial x^n}\right) \longleftrightarrow df = \frac{\partial f}{\partial x^1} dx + .. + \frac{\partial f}{\partial x^n} dx^n$$

so grad f corresponds to df.

Next suppose $\underline{a} \longleftrightarrow \alpha$, and that $\mathbb{R}^n$ has the standard orientation.

Then

$$\begin{aligned} {*d*}\alpha &= {*d*}(a_1 dx^1 + \ldots + a_n dx^n) \\ &= {*d}(a_1 dx^2 \wedge .. \wedge dx^n + \ldots + (-1)^{n-1} a_n dx^1 \wedge .. \wedge dx^{n-1}) \\ &= * \left(\left(\frac{\partial a_1}{\partial x^1} +, \ldots, + \frac{\partial a_n}{\partial x^n} \right) dx^1 \wedge .. \wedge dx^n \right) \\ &= \frac{\partial a_1}{\partial x^1} + .. + \frac{\partial a_n}{\partial x^n} \\ &= \text{div } \underline{a}. \end{aligned}$$

Thus div $\underline{a} = {*d*}\underline{\alpha}$.

From the expressions for grad and div it follows that we can write the Laplacian as $\nabla^2 f = {*d*}df$.

Finally, in the case where n = 3, $\mathbb{R}^3$ has the standard orientation and $\underline{a} \longleftrightarrow \alpha$ we have

$$*d\alpha = *d(a_1dx^1 + a_2dx^2 + a_3dx^3)$$

$$= *\left\{\left(\frac{\partial a_3}{\partial x^2} - \frac{\partial a_2}{\partial x^3}\right) dx^2 \wedge dx^3 + \left(\frac{\partial a_1}{\partial x^3} - \frac{\partial a_3}{\partial x^1}\right) dx^3 \wedge dx^2 + \left(\frac{\partial a_2}{\partial x^1} - \frac{\partial a_1}{\partial x^2}\right) dx^1 \wedge dx^2\right\}$$

$$= \left(\frac{\partial a_3}{\partial x^2} - \frac{\partial a_2}{\partial x^3}\right) dx^1 + \left(\frac{\partial a_1}{\partial x^3} - \frac{\partial a_3}{\partial x^1}\right) dx^2 + \left(\frac{\partial a_2}{\partial x^1} - \frac{\partial a_1}{\partial x^2}\right) dx^3$$

Thus curl $\underline{a} \longleftrightarrow *d\alpha$.

Notice also that the well-known indentities curl grad = 0, div curl = 0 correspond precisely to $d^2 = 0$ on 0-forms and 1-forms. For curl grad f = 0 corresponds to $*ddf = 0$ i.e. to $ddf = 0$, while div curl $\underline{a} = 0$ corresponds to $*d**d\alpha = 0$ i.e. to $dd\alpha = 0$.

As the reader has probably guessed by now there is a close relationship in 3-dimensions between the vector product $\underline{a} \times \underline{b}$ of vector fields and the exterior product $\alpha\wedge\beta$ of the corresponding 1-forms. In fact if $\underline{a} \longleftrightarrow \alpha$, $\underline{b} \longleftrightarrow \beta$ then

$$\underline{a} \times \underline{b} \longleftrightarrow *(\alpha\wedge\beta).$$

There is also a relationship involving the dot product, namely:

$$\underline{a}.\underline{b} = *(*\alpha\wedge\beta).$$

Using these relations one can easily establish the well-known identities of which the following is typical:

$$\mathrm{div}(\underline{a} \times \underline{b}) = \underline{b}.\ \mathrm{curl}\ \underline{a} - \underline{a}.\ \mathrm{curl}\ \underline{b}\ ;$$

for

$$\begin{aligned} *d*(*(\alpha\wedge\beta)) &= *d(\alpha\wedge\beta) \\ &= *(d\alpha\wedge\beta - \alpha\wedge d\beta). \\ &= *(*d\alpha\wedge*\beta - *\alpha\wedge*d\beta) \\ &= \underline{b}.\ \mathrm{curl}\ \underline{a} - \underline{a}.\ \mathrm{curl}\ \underline{b}. \end{aligned}$$

If we are working with a metric of the form $ds^2 = (h_1 dx^1)^2 + .. + (h_n dx^n)^2$, say, then the correspondence between vector fields and 1-forms is given by:

$$\underline{a} = (a^1, .. , a^n) \longleftrightarrow \alpha = a_1 h_1 dx^1 + \ldots + a_n h_n dx^n,$$

where $a_i = a^i$, and grad, div and curl may be computed in terms of these curvilinear coordinates using d, *d* and *d respectively.

<u>Examples</u>

(i) The Laplacian in polar coordinates. Here $ds^2 = dr^2 + (rd\theta)^2$. Then

$$\begin{aligned}\nabla^2 f &= *d*df.\\ &= *d*(f_r dr + f_\theta d\theta)\\ &= *d(-\frac{1}{r} f_\theta dr + rf_r d\theta)\\ &= *\{(\frac{1}{r} f_{\theta\theta} + (rf_r)_r) dr \wedge d\theta\}\\ &= \frac{1}{r}(rf_r)_r + \frac{1}{r^2} f_{\theta\theta}\end{aligned}$$

Thus $\nabla^2 f = \dfrac{\partial^2 f}{\partial r^2} + \dfrac{1}{r}\dfrac{\partial f}{\partial r} + \dfrac{1}{r^2}\dfrac{\partial^2 f}{\partial \theta^2}$.

(ii) If $\mathbb{R}^3$ has the metric $ds^2 = (h_1 dx^1)^2 + (h_2 dx^2)^2 + (h_3 dx^3)^2$ then grad $f = \left(\dfrac{1}{h_1}\dfrac{\partial f}{\partial x^1}, \dfrac{1}{h_2}\dfrac{\partial f}{\partial x^2}, \dfrac{1}{h_3}\dfrac{\partial f}{\partial x^3}\right)$,

while if $\underline{a} = (a^1, a^2, a^3) \longleftrightarrow \alpha = a_1 h_1 dx^1 + a_2 h_2 dx^2 + a_3 h_3 dx^3$, where $a_i = a^i$, then since $*h_1 dx^1 = h_2 dx^2 \wedge h_3 dx^3$ etc, we have

$$\begin{aligned}*d*\alpha &= *d*(a_1 h_1 dx^1 + a_2 h_2 dx^2 + a_3 h_3 dx^3)\\ &= *d(a_1 h_2 h_3 dx^2 \wedge dx^3 + a_2 h_3 h_1 dx^3 \wedge dx^1 + a_3 h_1 h_2 dx^1 \wedge dx^2)\\ &= *\left(\left(\frac{\partial(a_1 h_2 h_3)}{\partial x^1} + \frac{\partial(a_2 h_3 h_1)}{\partial x^2} + \frac{\partial(a_3 h_1 h_2)}{\partial x^3}\right) dx^1 \wedge dx^2 \wedge dx^3\right)\\ &= \frac{1}{h_1 h_2 h_3}\left(\frac{\partial(a_1 h_2 h_3)}{\partial x^1} + \frac{\partial(a_2 h_3 h_1)}{\partial x^2} + \frac{\partial(a_3 h_1 h_2)}{\partial x^3}\right)\end{aligned}$$

Thus $\operatorname{div} \underline{a} = \dfrac{1}{h_1h_2h_3}\left(\dfrac{\partial(a_1h_2h_3)}{\partial x^1} + \dfrac{\partial(a_2h_3h_1)}{\partial x^2} + \dfrac{\partial(a_3h_1h_2)}{\partial x^3}\right)$

Similar calculations can be performed for arbitrary metrics on $\mathbb{R}^n$.

6. STOKES' THEOREM

In this section we consider the integration of differential forms. Rather than setting out the details we concentrate on familiar examples and leave the interested reader to pursue the general theory elsewhere. (There are excellent treatments in [2,4] for example).

By an oriented p-dimensional submanifold M of $\mathbb{R}^3$ with boundary ∂M we shall mean:

(i) an oriented curve M with endpoints ∂M if $p = 1$;

(ii) an oriented surface with boundary curve(s) ∂M is $p = 2$;

(iii) an oriented 3-dimensional region M of $\mathbb{R}^3$ with boundary surface(s) ∂M if $p = 3$.

(We do not exclude the possibility that ∂M is empty as for example in the case where M is a simple closed curve or a closed surface.)

If α is a p-form defined o.. an oriented p-dimensional submanifold M of $\mathbb{R}^3$ then we define $\int_M \alpha$ to be

(i) the oriented line integral $\int_M f_1dx^1 + f_2dx^2 + f_3dx^3$ if $p = 1$ and $\alpha = f_1dx^1 + f_2dx^2 + f_3dx^3$,

(ii) the oriented surface integral $\int_M f_1dx^2dx^3 + f_2dx^3dx^1 + f_3dx^1dx^2$ if $p = 2$ and $\alpha = f_1dx^2 \wedge dx^3 + f_2dx^3 \wedge dx^1 + f_3dx^1 \wedge dx^2$,

(iii) the oriented volume integral $\int_M f\, dx^1dx^2dx^3$ if $p = 3$ and $\alpha = f\, dx^1 \wedge dx^2 \wedge dx^3$.

All these notions can be easily extended to higher dimensions so that we may define the integral of a p-form over an oriented p-dimensional submanifold of $\mathbb{R}^n$ (where $p \leq n$).

Then the generalized version of Stokes' Theorem, commonly referred to just as Stokes' Theorem, can be stated as follows.

Stokes' Theorem. Let M be an oriented p-dimensional submanifold of $\mathbb{R}^n$ with boundary ∂M and let α be a (p-1)-form defined on M. Then

$$\int_M d\alpha = \int_{\partial M} \alpha.$$

(If ∂M is empty then $\int_M d\alpha = 0$.)

This is not the most general version of the theorem nor the most useful for all purposes, but it is sufficiently general to indicate the elegance and usefulness of differential forms. There are several important remarks to make about this theorem.

1. In the special case where M is an interval [a, b] in the real line $\mathbb{R}^1$ and $\alpha = f(x)$ is a 0-form on [a, b] Stokes' Theorem reduces to the Fundamental Theorem of Calculus

$$\int_a^b f'(x)dx = f(b) - f(a).$$

Indeed, Stokes' Theorem is precisely the higher dimensional analogue of the Fundamental Theorem of Calculus.

2. The proof of Stokes' Theorem is very easy once the difficulties of defining what is meant by the integral of a p-form over a p-dimensional submanifold of $\mathbb{R}^n$ have been dealt with. It is simple consequence of the Fundamental Theorem of Calculus. See [4].

3. Green's Theorem, the Divergence and the classical Stokes Theorem are all special cases of Stokes' Theorem. The case of Green's Theorem has already been dealt with in §1. For the Divergence Theorem let us suppose that M is a 3-dimensional submanifold of $\mathbb{R}^3$ with boundary surface ∂M and let $\alpha = a_1 dx^2 \wedge dx^3 + a_2 dx^3 \wedge dx^1 + a_3 dx^1 \wedge dx^2$ be a 2-form defined on M. Then

$$d\alpha = \left(\frac{\partial a_1}{\partial x^1} + \frac{\partial a_2}{\partial x^2} + \frac{\partial a_3}{\partial x^3}\right) dx^1 \wedge dx^2 \wedge dx^3$$

and we have

$$\int_M \left(\frac{\partial a_1}{\partial x^1} + \frac{\partial a_2}{\partial x^2} + \frac{\partial a_3}{\partial x^3}\right) dx^1 \wedge dx^2 \wedge dx^3$$

$$= \int_{\partial M} (a_1 dx^2 \wedge dx^3 + a_2 dx^3 \wedge dx^1 + a_3 dx^1 \wedge dx^2).$$

This is more usually written as

$$\int_M \text{div } \underline{a} \;\, dV = \int_{\partial M} \underline{a}.d\underline{S}, \text{ where } \underline{a} = (a_1, a_2, a_3),$$

which is the Divergence Theorem.

For the classical Stokes Theorem let M be an oriented surface in $\mathbb{R}^3$ with boundary curve(s) ∂M and let $\alpha = a_1 dx^1 + a_2 dx^2 + a_3 dx^3$ be a 1-form defined on M. Then

$$d\alpha = \left(\frac{\partial a_3}{\partial x^2} - \frac{\partial a_2}{\partial x^3}\right) dx^2 \wedge dx^3 + \left(\frac{\partial a_1}{\partial x^3} - \frac{\partial a_3}{\partial x^1}\right) dx^3 \wedge dx^1$$

$$+ \left(\frac{\partial a_2}{\partial x^1} - \frac{\partial a_1}{\partial x^2}\right) dx^1 \wedge dx^2.$$

Then we have

$$\int_M \left(\left(\frac{\partial a_3}{\partial x^2} - \frac{\partial a_2}{\partial x^3}\right) dx^2 \wedge dx^3 + \left(\frac{\partial a_1}{\partial x^3} - \frac{\partial a_3}{\partial x^1}\right) dx^3 \wedge dx^1\right.$$

$$\left. + \left(\frac{\partial a_2}{\partial x^1} - \frac{\partial a_1}{\partial x^2}\right) dx^1 \wedge dx^2\right) = \int_{\partial M} (a_1 dx^1 + a_2 dx^2 + a_3 dx^3).$$

This is more usually written as

$$\int_M \text{curl } \underline{a}.d\underline{S} = \int_{\partial M} \underline{a}.d\underline{s}, \text{ where } \underline{a} = (a_1, a_2, a_3),$$

which is the classical Stokes Theorem.

4. Stokes' Theorem is also the foundation stone on which rests the use of differential forms for studying topological properties of differentiable manifolds (c.f. the earlier remarks). For more details see [5].

7. DIFFERENTIAL FORMS AS ALTERNATING MULTILINEAR FUNCTIONS

We have still not yet explained what differential forms are. To do so properly would require more space than is available here so we merely give a brief indication. Further details can be found in [2,4,5].

If $\underline{a}_1 = (a_{11}, .., a_{1n}), ..., \underline{a}_n = (a_{n1}, .., a_{nn})$ are vectors in $\mathbb{R}^n$ then $V(\underline{a}_1, ..., \underline{a}_n) = \det(a_{ij})$ is a function of the n vectors $\underline{a}_1, .., \underline{a}_n$ which is linear in each argument, and alternating in the sense that interchanging any pair of vectors $\underline{a}_i, \underline{a}_j$ changes the sign of $V(\underline{a}_1, .., \underline{a}_n)$ since this corresponds to interchanging two rows in the determinant. Thus $V(\underline{a}_1, .., \underline{a}_n)$ is an alternating multilinear function of n vectors. Furthermore, if $\mathbb{R}^n$ is given its standard euclidean metric, $|V(\underline{a}_1, .., \underline{a}_n)|$ is the volume of the parallelepiped $\Pi(\underline{a}_1, .., \underline{a}_n)$ determined by $\underline{a}_1, .., \underline{a}_n$, and $V(\underline{a}_1, .., \underline{a}_n)$ is positive or negative (assuming $\underline{a}_1, .., \underline{a}_n$ are linearly independent) according as the orientation of $\mathbb{R}^n$ determined by $\underline{a}_1, .., \underline{a}_n$ is the standard one or its opposite. Thus we may call $V(\underline{a}_1, .., \underline{a}_n)$ the "oriented" volume of $(\underline{a}_1, .., \underline{a}_n)$. In terms of the coordinate functions we have, in the case of the plane $\mathbb{R}^2$ for example,

$$V(\underline{a}_1, \underline{a}_2) = x^1(\underline{a}_1)x^2(\underline{a}_2) - x^2(\underline{a}_1)x^1(\underline{a}_2)$$

and we write this as

$$V(\underline{a}_1, \underline{a}_2) = 2(x^1 \wedge x^2)(\underline{a}_1, \underline{a}_2).$$

Similarly in the case of $\mathbb{R}^n$ we have

$$V(\underline{a}_1, .., \underline{a}_n) = \sum_{\sigma} (\text{sign } \sigma)\, x^{\sigma(1)}(\underline{a}_1) \ldots x^{\sigma(n)}(\underline{a}_n),$$

where the summation is over all permutations σ of 1, .., n, and we write this as

$$V(\underline{a}_1, .., \underline{a}_n) = n!(x^1 \wedge .. \wedge x^n)(\underline{a}_1, .., \underline{a}_n).$$

Next we recall that if f is a real-valued function defined on $\mathbb{R}^n$ then the differential df is a linear function which approximates f at each point in the sense that

$$f(\underline{a} + \underline{h}) - f(\underline{a}) = df_{\underline{a}}(h) + R(\underline{h})$$

where $df_{\underline{a}}(h)$ is a linear function of $\underline{h}$ and $\lim_{\underline{h} \to \underline{0}} \frac{R(h)}{|\underline{h}|} = \underline{0}$.

If $\underline{e}$ is a unit vector then $df_{\underline{a}}(\underline{e})$ is the rate of change of f at $\underline{a}$ in the direction $\underline{e}$ so that, for example, if $\underline{e}_i = (0, \ldots, 0,1,0, \ldots 0)$ is the i-th basis vector of $\mathbb{R}^n$, $df_{\underline{a}}(\underline{e}_i)x = \frac{\partial f}{\partial x^i}(a)$.

In considering differential forms as integrands of multiple integrals as suggested in the introduction these two notions are combined. A <u>differential p-form</u> α on $\mathbb{R}^n$ is an alternating multilinear function of p vector fields on $\mathbb{R}^n$ which assigns to p vector fields $\underline{a}_1, \ldots, \underline{a}_n$ on $\mathbb{R}^n$ the real valued function $\alpha(\underline{a}_1, \ldots, \underline{a}_n)$. In particular if we define

$$dx^i \wedge dx^j(\underline{a}_1, \underline{a}_2) = \frac{1}{2!}(dx^i(\underline{a}_1)dx^j(\underline{a}_2) - dx^j(\underline{a}_1)\, dx^i(\underline{a}_2)),$$

and more generally

$$dx^{i_1} \wedge \ldots \wedge dx^{i_p}(\underline{a}_1, \ldots, \underline{a}_p) = \frac{1}{p!}\sum_\sigma (\text{sign } \sigma) dx^{i_{\sigma(1)}}(\underline{a}_1) \ldots dx^{i_{\sigma(p)}}(\underline{a}_p),$$

then it is not hard to see that every p-form α may be written in the form $\alpha = f_{i_1 \ldots i_p} dx^{i_1} \wedge \ldots \wedge dx^{i_p}$. Furthermore, if α is a p-form and β is a q-form then $\alpha \wedge \beta$ is the (p+q)-form defined by

$$(\alpha \wedge \beta)(\underline{a}_1, \ldots, \underline{a}_{p+q})$$

$$= \frac{1}{(p+q)!}\sum_\sigma (\text{sign } \sigma)\alpha(\underline{a}_{\sigma(1)}, \ldots, \underline{a}_{\sigma(p)})\beta(\underline{a}_{\sigma(p+1)}, \ldots, \underline{a}_{\sigma(p+q)}).$$

The properties of differential forms mentioned in earlier sections and the role of differential forms in multiple integrals are now easy to deduce.

REFERENCES

[1] Cartan, É., *Oeuvres complètes*, Gauthier-Villars, Paris, 1952-55. Reprinted by Springer, 1984.

[2] Cartan, H., *Formes Differentielles*, Hermann, Paris, 1967.

[3] Flanders, H., *Differential Forms with Applications to the Physical Sciences*, Academic Press, 1963.

[4] Spivak, M., *Calculus on Manifolds*, Benjamin, New York, 1965.

[5] Spivak, M., *A Comprehensive Introduction to Differential Geometry I*, Publish or Perish, 1979.

DIFFERENTIAL FORMS IN THE THEORY OF SURFACES

L.M. Woodward
(University of Durham)

1. INTRODUCTION

After introducing and developing the notion of differential forms in the context of Pfaffian systems Élie Cartan turned his attention to the application of the exterior differential calculus to differential geometry. Coupled with the idea of moving frames, which had previously been used by Darboux, differential forms were developed by Cartan into a very powerful and elegant tool which affords not only a clarity in the theoretical development of the differential geometry of surfaces and higher dimensional manifolds but also a great economy in calculation. His ideas, which were extended by de Rham and Hodge in the 1930s and 1940s, have provided constant stimulus ever since and still form the basis for a large part of current research in differential geometry.

In order to understand these ideas it is necessary first to grasp the role of moving orthonormal frames. This is already familiar from the theory of space curves. Recall that if α is a curve in 3-space $\mathbb{R}^3$ parametrized by arc length then at each point of α the unit tangent vector $\underline{t}_\alpha = \alpha'$ is defined. If furthermore α'' is non-zero, then we write $\alpha'' = \kappa_\alpha \underline{n}_\alpha$, where $\kappa_\alpha = |\alpha''|$ is the curvature of α and $\underline{n}_\alpha$ is a unit vector called the principal normal of α. Finally we define the unit binormal $\underline{b}_\alpha$ of α by $\underline{b}_\alpha = \underline{t}_\alpha \times \underline{n}_\alpha$. Thus at each point of α we have an orthonormal frame $\underline{t}_\alpha$, $\underline{n}_\alpha$, $\underline{b}_\alpha$ and the whole of the local differential geometry of the curve is described by the rate of change of this frame as we move along the curve.

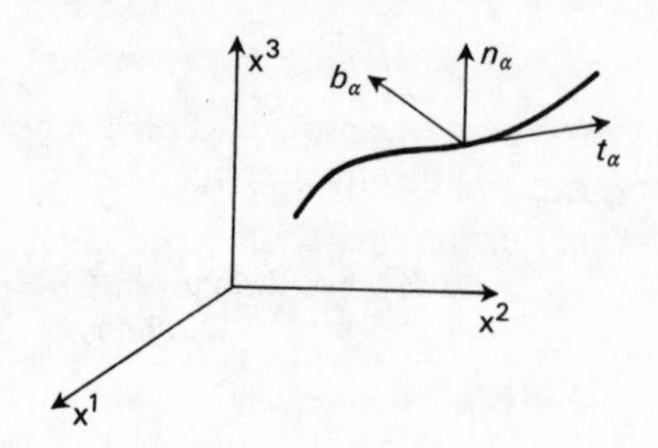

Fig. 1. The moving frame

Indeed as is well-known $\underline{b}'_\alpha = -\tau_\alpha \underline{n}_\alpha$, where τ_α is called the torsion of α, we have the Serret-Frenet formulae

$$\begin{aligned} \underline{t}'_\alpha &= \kappa_\alpha \underline{n}_\alpha \\ \underline{n}'_\alpha &= -\kappa_\alpha \underline{t}_\alpha + \tau_\alpha \underline{b}_\alpha \\ \underline{b}'_\alpha &= -\tau_\alpha \underline{n}_\alpha \end{aligned} \qquad (1.1),$$

and the Fundamental Theorem of the Local Theory of Curves says that the curve α is uniquely determined by $\kappa_\alpha, \tau_\alpha$ considered as functions of the arc length s. The Serret-Frenet Formulae (1.1) can also be written as follows: let A be the 3×3 matrix whose columns are $\underline{t}_\alpha$, $\underline{n}_\alpha$, $\underline{b}_\alpha$. Then (1.1) is equivalent to

$$A' = A \begin{pmatrix} 0 & -\kappa_\alpha & 0 \\ \kappa_\alpha & 0 & -\tau_\alpha \\ 0 & \tau_\alpha & 0 \end{pmatrix}$$

or equivalently

$$A^{-1} \quad dA = \begin{pmatrix} 0 & -\kappa_\alpha ds & 0 \\ \kappa_\alpha ds & 0 & -\tau_\alpha ds \\ 0 & \tau_\alpha ds & 0 \end{pmatrix} \qquad (1.2)$$

Notice that the matrix on the right hand side is skew-symmetric. This is essentially because the moving frame defines a curve in the space SO_3 of real orthogonal 3×3 matrices of determinant

+ 1 and from the identity $A(s)^t A(s) = I$ (where t denotes transpose) we have $(A^t)'A + A^tA' = 0$ so that $A^{-1}A'$ is skew-symmetric. Thus the geometry of a curve α in $\mathbb{R}^3$ is described by a curve in SO_3 (without any particular geometrical properties) and the Fundamental Theorem says, in effect, that this curve in SO_3 is sufficient to describe the original curve in $\mathbb{R}^3$ without specifying any integrability conditions. This is not particularly helpful in dealing with curves but will be useful later as a point of reference and comparison when discussing surfaces.

The classical approach to the local theory of surfaces is via coordinate patches. Let $\underline{x}(u^1, u^2) = (x^1(u^1,u^2), x^2(u^1,u^2), x^3(u^1,u^2))$ where (u^1,u^2) lies in some open set U in $\mathbb{R}^2$, define a coordinate patch on a surface S in $\mathbb{R}^3$.

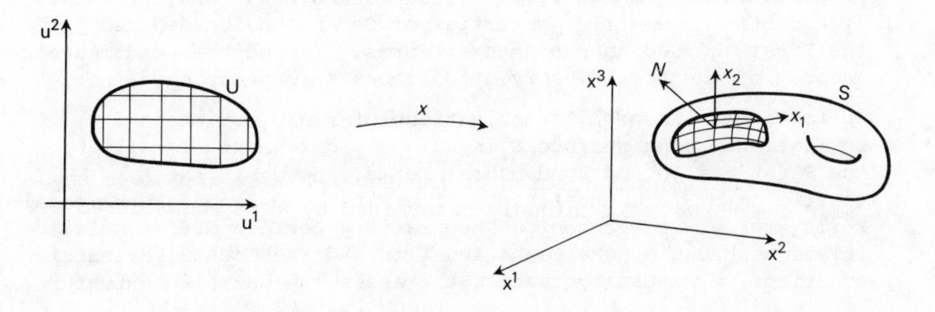

Fig. 2. A coordinate patch

Then the vectors $\underline{x}_1, \underline{x}_2$, $\left(\text{where } \underline{x}_j = \dfrac{\partial \underline{x}}{\partial u^j}\right)$ are the tangent vectors to the coordinate curves u_2 = constant, u_1 = constant respectively, and the metric, or First Fundamental Form, is defined by

$$I = ds^2 = d\underline{x}.d\underline{x} = E(du^1)^2 + 2F\, du^1 du^2 + G(du^2)^2,$$

where $E = \underline{x}_1 . \underline{x}_1$, $F = \underline{x}_1 . \underline{x}_2$, $G = \underline{x}_2 . \underline{x}_2$. This is often written more concisely in tensor notation with summation convention as $ds^2 = g_{ij}\, du^i du^j$ where $g_{ij} = \underline{x}_i . \underline{x}_j$. The unit normal $\underline{N}$ is defined by $\underline{N} = \dfrac{\underline{x}_1 \times \underline{x}_2}{|\underline{x}_1 \times \underline{x}_2|}$ and the curvature of the surface S is described by the rate of change of $\underline{N}$, that is to say by the differential $d\underline{N}$. We recall that if $\underline{e}$ is a unit tangent vector to S at p then $d\underline{N}_p(e)$ is the rate of change of $\underline{N}$ at p in the direction $\underline{e}$. The actual analysis of the differential $d\underline{N}$ is usually achieved via the Second Fundamental Form

$$II = -d\underline{N}.d\underline{x} = L(du^1)^2 + 2M\, du^1 du^2 + N(du^2)^2,$$

where $L = \underline{N}.\underline{x}_{11}$, $M = \underline{N}.\underline{x}_{12}$, $N = \underline{N}.\underline{x}_{22}$, $\underline{x}_{ij} = \dfrac{\partial^2 \underline{x}}{\partial u^i \partial u^j}$; or

$II = L_{ij}\, du^i du^j$, where $L_{ij} = \underline{N}.\underline{x}_{ij}$, in tensor notation. Indeed, given the First Fundamental Form, because of some beautiful symmetry properties of $d\underline{N}$, the differential $d\underline{N}$ and the Second Fundamental Form uniquely determine each other. All the local differential geometric properties of S can then be deduced from the First and Second Fundamental Forms. Indeed the Fundamental Theorem of the Local Theory of Surfaces in essence says that, up to isometries of $\mathbb{R}^3$ (combinations of rigid motions and reflections), the surface S in uniquely determined locally by the First and Second Fundamental Forms. However, the coefficients E,F,G and L,M,N of the First and Second Fundamental Forms are not independent. They satisfy certain differential relations known as the Gauss equation and the Codazzi-Mainardi equations. The Gauss equation expresses the Gaussian curvature

$K = \dfrac{LN-M^2}{EG-F^2}$ entirely in terms of E,F,G and their derivatives

while the Codazzi-Mainardi equations appear at first sight more mysterious and are more difficult to describe. In the special case of orthogonal coordinates where F = O the equations reduce to

$$K = -\frac{1}{2\sqrt{EG}}\left\{\left(\frac{E_2}{\sqrt{EG}}\right)_2 + \left(\frac{G_1}{\sqrt{EG}}\right)_1\right\} \quad \text{(Gauss equation)} \tag{1.3}$$

$$L_2 - M_1 = \frac{1}{2}\left(\frac{L}{E} + \frac{N}{G}\right)E_2 - \frac{1}{2}M\left(\frac{E_1}{E} + \frac{G_1}{G}\right) \tag{1.4}$$

(Codazzi-Mainardi equations)

$$N_1 - M_2 = \frac{1}{2}\left(\frac{L}{E} + \frac{N}{G}\right)G_1 - \frac{1}{2}M\left(\frac{E_2}{E} + \frac{G_2}{G}\right) \tag{1.5}$$

These equations and the expressions for the Christoffel symbols Γ^k_{ij}, which are implicitly involved here, are derived by using the fact that at each point of the coordinate neighbourhood on the surface S we have a frame $\underline{x}_1$, $\underline{x}_2$, $\underline{N}$. The Christoffel symbols, for example, are defined by

$$\underline{x}_{ij} = \Gamma^k_{ij}\,\underline{x}_k + L_{ij}\,\underline{N} \tag{1.6}$$

(where $L_{11} = L$, $L_{12} = L_{21} = M$, $L_{22} = N$) and the Gauss and Codazzi-Mainardi equations are obtained by writing out the identities

$$(\underline{x}_{11})_2 = (\underline{x}_{12})_1, (\underline{x}_{22})_1 = (\underline{x}_{12})_2, \underline{N}_{12} = \underline{N}_{21}$$

in terms of their components with respect to $\underline{x}_1$, $\underline{x}_2$, $\underline{N}$.

This is all well-known and perfectly adequate as a tool for dealing with the local differential geometry of surfaces. Why, then, should one attempt to do things differently? The main reason is this. Unlike the sensible course we adopt for curves, where we deal only with orthonormal frames, here we are using frames $\underline{x}_1$, $\underline{x}_2$, $\underline{N}$ which in general are not even orthogonal. This leads to more complicated expressions than are really necessary, complicates and lengthens calculations, and obscures many of the geometrical aspects of what we are doing. Given carte blanche few of us would choose to do Euclidean geometry with oblique axes and yet that is, in effect, what we are doing here. Cartan's contribution is to use orthonormal frames and then to further simplify matters by using the exterior differential calculus.

Briefly, his approach is as follows. As before, we assume that we are given a coordinate neighbourhood $\underline{x}(u^1,u^2) = (x^1(u^1,u^2), x^2(u^1,u^2), x^3(u^1,u^2))$. We then choose an orthonormal moving frame $\underline{e}_1$, $\underline{e}_2$, $\underline{e}_3$ in this coordinate neighbourhood with $\underline{e}_3$ normal to the surface S. Thus $\underline{e}_1$, $\underline{e}_2$, $\underline{e}_3$ are vector valued functions of (u^1, u^2) with $\underline{e}_i.\underline{e}_j = \delta_{ij}$ and $\underline{e}_3$ normal to S. For example we could choose $\underline{e}_1$, $\underline{e}_2$ to be obtained from $\underline{x}_1$, $\underline{x}_2$ by Gram-Schmidt orthonormalization and let $\underline{e}_3 = \underline{N}$. Then we have

$$d\underline{x} = \theta^i \underline{e}_i \quad (= \theta^1 \underline{e}_1 + \theta^2 \underline{e}_2 + \theta^3 \underline{e}_3) \tag{1.6}$$

where $\theta^1, \theta^2, \theta^3$ are 1-forms with $\theta^3 = 0$.

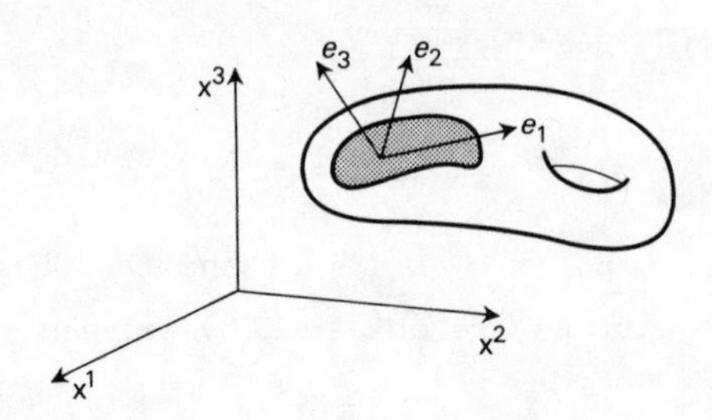

Fig. 3. An orthonormal frame

The First Fundamental Form is given by

$$I = ds^2 = d\underline{x}.d\underline{x} = (\theta^1)^2 + (\theta^2)^2 \tag{1.7}$$

Also

$$d\underline{e}_i = \omega_i^j \, \underline{e}_j \tag{1.8}$$

for some 1-forms ω_i^j and from the condition $\underline{e}_i . \underline{e}_j = \delta_{ij}$ we deduce that $\omega_i^j + \omega_j^i = 0$. Thus if we let A be the 3×3 real orthogonal matrix with columns $\underline{e}_1, \underline{e}_2, \underline{e}_3$ then (1.8) can be written

$$A^{-1} \, dA = (\omega_j^i) = \begin{pmatrix} \omega_1^1 & \omega_2^1 & \omega_3^1 \\ \omega_1^2 & \omega_2^2 & \omega_3^2 \\ \omega_1^3 & \omega_2^3 & \omega_3^3 \end{pmatrix}$$

and the matrix (ω_j^i) is skew-symmetric. This means that there are just 3 components $\omega_1^2, \omega_1^3, \omega_2^3$ involved in describing the rate of change of the moving frame instead of the two involved with space curves. (Roughly speaking these 3 components correspond to geodesic curvature, normal curvature and geodesic torsion - see §2). Furthermore from the fact that $d^2 = 0$ we deduce the <u>structure equations</u>:

$$d\theta^i + \omega_j^i \wedge \theta^j = 0, \quad i = 1,2,3$$

and $$d\omega^i_j + \omega^i_k \wedge \omega^k_j = 0, \quad i,j = 1,2,3$$

The second of these sets of equations can be written out explicitly, using the skew-symmetry of the matrix (ω^i_j), as

$$d\omega^2_1 + \omega^2_3 \wedge \omega^3_1 = 0 \qquad \text{Gauss equation}$$

$$d\omega^2_1 + \omega^3_2 \wedge \omega^2_1 = 0$$

$$d\omega^3_2 + \omega^3_1 \wedge \omega^1_2 = 0$$

Codazzi-Mainardi equations

(The equivalence of these equations with the Gauss and Codazzi-Mainardi equations mentioned earlier is discussed in §4.) Furthermore, from the Frobenius Integrability Theory for Pfaffian systems [3] these are precisely the integrability conditions required to ensure the existence of an orthonormal frame field $\underline{e}_1$, $\underline{e}_2$, $\underline{e}_3$ such that $d\underline{e}_i = \omega^j_i \, \underline{e}_j$.

The first set of equations may be written out as:

$$d\theta^1 + \omega^1_2 \wedge \theta^2 = 0$$

$$d\theta^2 + \omega^2_1 \wedge \theta^1 = 0$$

$$\omega^3_1 \wedge \theta^1 + \omega^3_2 \wedge \theta^2 = 0 \qquad \text{(Symmetry equation)}$$

and these equations, by the Frobenius Integrability Theorem, are precisely the integrability conditions required to ensure that the equations $d\underline{x} = \theta^1 \underline{e}_1 + \theta^2 \underline{e}_2$ can be integrated up to give a coordinate patch $\underline{x}(u^1, u^2)$ for a surface with First Fundamental Form $I = (\theta^1)^2 + (\theta^2)^2$.

Three features of this approach should be recognised. The first is that the expressions involved in the equations are much simpler than in the classical treatment. Secondly, assuming familiarity with the Frobenius Integrability Theorem, the significance of the Gauss and Codazzi-Mainardi equations is immediately apparent. Finally the analogy of the structure equations with the Serret-Frenet formulae is clear. Indeed Cartan developed this aspect of moving frames much further. He viewed the structure equations as arising from equations describing the structure of the Lie group SO_3. A curve or surface in $\mathbb{R}^3$ with a moving frame corresponds to a map of the curve or surface into SO_3 and via this map the structure equations for SO_3 are translated into the Serret-Frenet formulae for the curve or the structure equations for the surface. It is clear though not relevant to our particular interests here that all this generalizes very simply to the case of surfaces and higher dimensional manifolds in $\mathbb{R}^n$ where SO_3 is replaced by the Lie group SO_n of real orthogonal $n \times n$ matrices of determinant 1.

Whether differential forms will prove to be a useful tool in computer graphics or computer aided geometric design remains to be seen. For the case of curves there is nothing to be gained since the only non-zero differential forms on a curve are 0-forms (functions) and 1-forms (essentially differentials of functions in this case). However for surfaces the use of differential forms may turn out to be useful since calculations using forms involve fewer symbols and are in general simpler to perform thus reducing the possibility of errors occurring. If, as seems not unlikely, higher dimensional considerations arise then the usefulness of differential forms will become more apparent.

2. MOVING FRAMES AND THE DERIVATION OF THE STRUCTURE EQUATIONS

Let $\underline{x}(u^1,u^2) = (x^1(u^1,u^2), x^2(u^1,u^2), x^3(u^1,u^2))$ be a coordinate neighbourhood on a surface S where (u^1,u^2) belongs to an open set U in $\mathbb{R}^2$, and let $\underline{e}_1$, $\underline{e}_2$, $\underline{e}_3$ be a moving orthonormal frame on the coordinate neighbourhood with $\underline{e}_3$ normal to the surface. Then $d\underline{x}$ is a vector valued 1-form: in fact writing $\underline{x}$ as a column vector we have

$$d\underline{x} = \begin{pmatrix} dx^1 \\ dx^2 \\ dx^3 \end{pmatrix} = \begin{pmatrix} \dfrac{\partial x^1}{\partial u^1} du^1 + \dfrac{\partial x^1}{\partial u^2} du^2 \\[2ex] \dfrac{\partial x^2}{\partial u^1} du^1 + \dfrac{\partial x^2}{\partial u^2} du^2 \\[2ex] \dfrac{\partial x^3}{\partial u^1} du^1 + \dfrac{\partial x^3}{\partial u^2} du^2 \end{pmatrix}$$

However we are not concerned with the components of $d\underline{x}$ with respect to the standard basis of $\mathbb{R}^3$ but with respect to the moving frame $\underline{e}_1, \underline{e}_2, \underline{e}_3$. We have

$$d\underline{x} = \theta^i \underline{e}_i \tag{2.1}$$

for some choice of 1-forms $\theta^1, \theta^2, \theta^3$, and $\theta^3 = 0$ since $d\underline{x} = \underline{x}_1\, du^1 + \underline{x}_2\, du^2$ is tangential to the surface. The First Fundamental Form, or metric, is then given by

$$I = ds^2 = d\underline{x}.d\underline{x} = (\theta^1)^2 + (\theta^2)^2 \tag{2.2}$$

and the element of area, or area form is $\theta^1 \wedge \theta^2$.

Similarly there exist 1-forms ω^j_i such that

$$d\underline{e}_i = \omega^j_i\, \underline{e}_j\,, \quad i = 1,2,3. \tag{2.3}$$

Since $\underline{e}_i . \underline{e}_j = \delta_{ij}$ we have

$$d\underline{e}_i . \underline{e}_j + \underline{e}_i . d\underline{e}_j = 0$$

and hence

$$\omega^j_i + \omega^i_j = 0 \tag{2.4}$$

Thus the matrix (ω^j_i) of 1-forms is skew-symmetric. In particular the whole matrix (ω^j_i) is known once $\omega^2_1, \omega^3_1, \omega^3_2$, are known.

Example: Let S be the torus of revolution obtained by rotating the circle

$$(x - a)^2 + (z) = b^2, \ y = 0 \quad (a > b > 0)$$

around the z-axis. Consider the coordinate neighbourhood

$$\underline{x}(u,v) = ((a+b \cos v)\cos u, \ (a+b \cos v)\sin u, \ b \sin v)$$

where $0 < u < 2\pi$, $0 < v < 2\pi$. Here

$$\underline{x}_1 = (a+b \cos v)\ (-\sin u, \cos u, 0).$$

$$\underline{x}_2 = b(-\sin v \cos u, -\sin v \sin u, \cos v)$$

so that we may choose $\underline{e}_1 = (\sin u, \cos u, 0)$,

$$\underline{e}_2 = (-\cos u \sin v, - \sin u \sin v, \cos v).$$

$\underline{e}_3$ = (cos u cos v, sin u cos v, sin v) for example.

Then as $d\underline{x} = \underline{x}_u\, du + \underline{x}_v\, dv = \theta^1\, \underline{e}_1 + \theta^2\, \underline{e}_2$ we have $\theta^1 =$ $(a+b \cos v)\, du$, $\theta^2 = b\, dv$. In this case the metric is given by $ds^2 = (\theta^1)^2 + (\theta^2)^2 = (a+b \cos v)^2\, (du)^2 + b^2 (dv)^2$.

Also
$$d\underline{e}_1 = -(\cos u, \sin u, 0)\, du$$
$$= \sin v\, du\, \underline{e}_2 - \cos u\, du\, \underline{e}_3$$
$$d\underline{e}_2 = (\sin u \sin v, - \cos u \sin v, 0)\, du$$
$$- (\cos u \cos v, \sin u \cos v, \sin v)\, dv$$
$$= - \sin v\, du\, \underline{e}_1 - dv\, \underline{e}_3.$$
$$d\underline{e}_3 = (-\sin u \cos v, \cos u \cos v, 0)\, du$$
$$+ (-\cos u \sin v, - \sin u \sin v, \cos v)\, dv$$
$$= \cos v\, du\, \underline{e}_1 + dv\, \underline{e}_2.$$

Thus
$$\omega_1^2 = \sin v\, du = - \omega_2^1$$
$$\omega_1^3 = - \cos v\, du = - \omega_3^1$$
$$\omega_2^3 = - dv = - \omega_3^2 .$$

Applying the exterior differential operator d to equation (2.1) and using the fact that $d^2 = 0$ we have

$$\underline{0} = d(d\underline{x})$$
$$= d(\theta^i\, \underline{e}_i)$$
$$= d\theta^i\, \underline{e}_i - \theta^i \wedge d\underline{e}_i$$
$$= d\theta^i\, \underline{e}_i - \theta^j \wedge d\underline{e}_j$$

$$= (d\theta^i - \theta^j \wedge \omega^i_j)\, \underline{e}_i$$

by (2.3). Thus $d\theta^i - \theta^j \wedge \omega^i_j = 0$ or, equivalently,

$$d\theta^i + \omega^i_j \wedge \theta^j = 0\ , \quad i = 1,2,3. \tag{2.5}$$

Similarly

$$\underline{0} = d(d\, \underline{e}_i)$$

$$= d(\omega^j_i\, \underline{e}_j)$$

by (2.3)

$$= (d\omega^j_i - \omega^k_i \wedge \omega^j_k)\, \underline{e}_j,$$

so that

$$d\omega^j_i + \omega^j_k \wedge \omega^k_i = 0 \tag{2.6}$$

Using the fact that $\theta^3 = 0$ and that $\omega^j_i + \omega^i_j = 0$ we can write (2.5) and (2.6) as

$$d\theta^1 + \omega^1_2 \wedge \theta^2 = 0 \tag{2.7}$$

$$d\theta^2 + \omega^2_1 \wedge \theta^1 = 0 \tag{2.8}$$

$$\omega^3_1 \wedge \theta^1 + \omega^3_2 \wedge \theta^2 = 0 \qquad \text{Symmetry equation} \tag{2.9}$$

$$d\omega^2_1 + \omega^2_3 \wedge \omega^3_1 = 0 \qquad \text{Gauss equation} \tag{2.10}$$

$$d\omega^3_1 + \omega^3_2 \wedge \omega^2_1 = 0 \tag{2.11}$$

$$d\omega^3_2 + \omega^3_1 \wedge \omega^1_2 = 0 \tag{2.12}$$

Codazzi-Mainardi equations (2.11), (2.12)

A geometrical interpretation of the forms ω^2_1, ω^3_1, ω^3_2 may be given by relating them to geodesic curvature, normal curvature and geodesic torsion. Recall that if α is a curve on S, parametrized by arc length, then at each point of α we have an orthonormal frame α', $\underline{N} \times \alpha'$, $\underline{N}$ and that

$$\alpha'' = \kappa_g\, \underline{N} \times \alpha' + \kappa_n\, \underline{N} \tag{2.13}$$

$$(\underline{N} \times \alpha')' = -\kappa_g\, \alpha' + \tau_g\, \underline{N} \tag{2.14}$$

$$\underline{N}' = -\kappa_n \alpha' - \tau_g \underline{N} \times a' \tag{2.15}$$

where κ_g, κ_n and τ_g are the geodesic curvature, normal curvature and geodesic torsion of α. We also recall that κ_n and τ_g depend only on the surface and the tangent vector α' at each point while κ_g measures the failure of the curve to be a geodesic. See [5].

Let $\underline{e}_1$, $\underline{e}_2$, $\underline{e}_3$ be a moving frame in a coordinate neighbourhood with $\underline{e}_3 = \underline{N}$ and let

$$\underline{e} = \alpha' = \cos\phi\, \underline{e}_1 + \sin\theta\, \underline{e}_2 \tag{2.16}$$

so that

$$\underline{N} \times \alpha' = -\sin\phi\, \underline{e}_1 + \cos\phi\, \underline{e}_2 \tag{2.17}$$

Then

$$d\underline{e} = (-\sin\phi\, \underline{e}_1 + \cos\phi\, \underline{e}_2)d\phi + \cos\phi\, d\underline{e}_1 + \sin\phi\, d\underline{e}_2$$

$$= (d\phi + \omega_1^2)\ (-\sin\phi\, \underline{e}_1 + \cos\phi\, \underline{e}_2) +$$

$$(\cos\phi\, \omega_1^3 + \sin\phi\, \omega_2^3)\underline{e}_3.$$

Since $\alpha'' = d\underline{e}(\underline{e})$ we have

$$\kappa_g = \frac{d\phi}{ds} + \omega_1^2(e) \tag{2.18}$$

$$\kappa_n = \cos\phi\, \omega_1^3(e) + \sin\phi\, \omega_2^3(\underline{e}). \tag{2.19}$$

Also since

$$\underline{N}' = d\underline{e}_3(e) = -\omega_1^3(\underline{e})\underline{e}_1 - \omega_2^3(\underline{e})\underline{e}_2$$

and $\underline{e}_1 = \cos\phi\, \alpha' - \sin\phi\, \underline{N} \times \alpha'$, $\underline{e}_2 = \sin\phi\, \alpha' + \cos\phi\, \underline{N} \times \alpha'$ we have

$$\kappa_n \alpha' + \tau_g \underline{N} \times \alpha' = (\cos\phi\, \omega_1^3(\underline{e}) + \sin\phi\, \omega_2^3(\underline{e}))\alpha'$$

$$+ (-\sin\phi\, \omega_1^3(\underline{e}) + \cos\phi\, \omega_2^3(\underline{e}))\underline{N} \times \alpha',$$

so

$$\tau_g = -\sin\phi\ \omega_1^3(\underline{e}) + \cos\phi\ \omega_2^3(\underline{e}). \tag{2.20}$$

Thus

$$\omega_1^2(\underline{e}) = \kappa_g - \frac{d\phi}{ds} \tag{2.21}$$

$$\omega_1^3(\underline{e}) = \cos\phi\ \kappa_n - \sin\phi\ \tau_g \tag{2.22}$$

$$\omega_2^3(\underline{e}) = \sin\phi\ \kappa_n + \cos\phi\ \tau_g \tag{2.23}$$

In particular if the moving frame $\underline{e}_1, \underline{e}_2, \underline{e}_3$ is adapted to the curve α so that $\underline{e}_1 = \underline{e} = \alpha'$ then we have $\omega_1^2(\underline{e}) = \kappa_g$, $\omega_1^3(\underline{e}) = \kappa_n$, $\omega_2^3(\underline{e}) = \tau_g$.

3. THE SECOND FUNDAMENTAL FORM

We recall that the Second Fundamental Form is defined by $\text{II} = -d\underline{e}_3.d\underline{x}$, and that $d\underline{x} = \theta^1\underline{e}_1 + \theta^2\underline{e}_2$, $d\underline{e}_3 = \omega_3^1\underline{e}_1 + \omega_3^2\underline{e}_2$. Let us write $\omega_i^3 = h_{i1}\theta^1 + h_{i2}\theta^2$ so that $\omega_i^3(\underline{e}_j) = h_{ij}$ (since $\theta^i(\underline{e}_j) = \delta_j^i$).

Then

$$-d\underline{e}_3 = \omega_1^3\underline{e}_1 + \omega_2^3\underline{e}_2$$

and so

$$-d\underline{e}_3(\underline{e}_j) = \omega_1^3(\underline{e}_j)\underline{e}_1 + \omega_2^3(\underline{e}_j)\underline{e}_2$$

$$= h_{1j}\underline{e}_1 + h_{2j}\underline{e}_2.$$

Thus with respect to the basis $\underline{e}_1, \underline{e}_2$ of the tangent plane to S at each point of the coordinate neighbourhood the linear map $-d\underline{e}_3$ is represented by the matrix

$$\begin{pmatrix} h_{11} & h_{12} \\ h_{21} & h_{22} \end{pmatrix}$$

From the symmetry equation $\omega_1^3 \wedge \theta^1 + \omega_2^3 \wedge \theta^2 = 0$ we have

$$(-h_{12} + h_{21})\, \theta^1 \wedge \theta^2 = 0$$

and hence

$$h_{12} = h_{21}. \tag{3.1}$$

Thus the matrix (h_{ij}) is symmetric and hence diagonalizable by a suitable rotation of the coordinates. In particular the eigenvalues κ_1, κ_2 of $-d\underline{e}_3$, known as the principal curvatures, are real and the <u>mean curvature</u> H and <u>Gaussian curvature</u> K are given by

$$H = \frac{1}{2}(\kappa_1 + \kappa_2) = \frac{1}{2}\,\mathrm{tr}\,(-d\underline{e}_3) = \frac{1}{2}(h_{11} + h_{22}) \tag{3.2}$$

$$K = \kappa_1\kappa_2 = \det\,(-d\underline{e}_3) = h_{11}h_{22} - h_{12}^2 \tag{3.3}$$

Furthermore the Second Fundamental Form is given by

$$\begin{aligned} II &= -d\underline{e}_3 . d\underline{x} \\ &= (\omega_1^3\underline{e}_1 + \omega_2^3\underline{e}_2).(\theta^1\underline{e}_1 + \theta^2\underline{e}_2) \\ &= \omega_1^3\theta^1 + \omega_2^3\theta^2 \\ &= h_{11}(\theta^1)^2 + 2h_{12}\,\theta^1\theta^2 + h_{22}(\theta^2)^2 \end{aligned}$$

i.e.

$$II = h_{11}(\theta^1)^2 + 2h_{12}\,\theta^1\theta^2 + h_{22}(\theta^2)^2 \tag{3.4}$$

If $\underline{e} = \cos\phi\, \underline{e}_1 + \sin\phi\, \underline{e}_2$ then

$$\begin{aligned} II(\underline{e}) &= -d\underline{e}_3(\underline{e}).e \\ &= (\omega_1^3(\underline{e})\underline{e}_1 + \omega_2^3(\underline{e})e_2).(\cos\phi\,\underline{e}_1 + \sin\phi\,\underline{e}_2) \\ &= \cos\phi\,\omega_1^3(\underline{e}) + \sin\phi\,\omega_2^3(\underline{e}) \\ &= \kappa_n(\underline{e}) \end{aligned}$$

the normal curvature in the direction $\underline{e}$. Also, since the critical values of the function $f(x,y) = ax^2 + 2bxy + cy^2$ subject to the condition $g(x,y) = x^2 + y^2 = 1$ are the eigenvalues of the matrix $\begin{pmatrix} a & b \\ b & c \end{pmatrix}$ it follows that the principal curvatures as defined above are the critical values of the normal curvature.

Note that changing $\underline{e}_3$ to $-\underline{e}_3$, which corresponds to a change of orientation, leaves K unchanged but changes H to -H. The mean curvature vector $\underline{H} = H\underline{e}_3$ is therefore independent of the choice of orientation. It is an easy matter using differential forms to show that $\underline{H} = \frac{1}{2} \nabla^2 \underline{x}$, where ∇^2 denotes the Laplacian. For

$$\begin{aligned}
\nabla^2 \underline{x} &= * \, d * d \, \underline{x} \\
&= * \, d * (\theta^1 \underline{e}_1 + \theta^2 \underline{e}_2) \\
&= * \, d \, (\theta^2 \underline{e}_1 - \theta^1 \underline{e}_2) \\
&= * \, \{d\theta^2 \underline{e}_1 - \theta^2 \wedge d\underline{e}_1 - d\theta^1 \underline{e}_2 + \theta^1 \wedge d\underline{e}_2\} \\
&= * \, \{d\theta^2 \underline{e}_1 - \theta^2 \wedge \omega_1^2 \underline{e}_2 - \theta^2 \wedge \omega_1^3 \underline{e}_3 \\
&\qquad - d\theta^1 \underline{e}_2 + \theta^1 \wedge \omega_2^1 \underline{e}_1 + \theta^1 \wedge \omega_2^3 \underline{e}_3\} \\
&= * \, \{\theta^1 \wedge \omega_2^3 - \theta^2 \wedge \omega_1^3\} \underline{e}_3 \quad \text{using (2.7) and (2.8)} \\
&= * \, \{h_{11} + h_{22}) \theta^1 \wedge \theta^2\} \underline{e}_3 \\
&= 2\underline{H}. \qquad (3.5)
\end{aligned}$$

Thus $\nabla^2 \underline{x} = 2\underline{H}$

In particular it follows that the surface is minimal (i.e. $\underline{H} = 0$) if and only if the components x^1, x^2, x^3 of $\underline{x}$ are harmonic functions of u^1, u^2. This is an important result which forms the basis for the Weierstrass representation of minimal surfaces by complex analytic functions, one of the most beautiful topics in the geometry of surfaces.

4. THE GAUSS AND CODAZZI-MAINARDI EQUATIONS

We first consider the Gauss equation. Since $\omega_i^3 = h_{ij}\theta^j$ we have

$$\begin{aligned} d\omega_1^2 &= -\omega_3^2 \wedge \omega_1^3 \\ &= -\omega_1^3 \wedge \omega_2^3 \\ &= -(h_{11}\theta^1 + h_{12}\theta^2) \wedge (h_{12}\theta^1 + h_{22}\theta^2) \\ &= -(h_{11}h_{22} - h_{12}^2)\theta^1 \wedge \theta^2 \end{aligned}$$

i.e. $$d\omega_1^2 = -K\theta^1 \wedge \theta^2. \tag{4.1}$$

This has several important consequencies. Firstly, consider $\underline{e}_3$ as a map of the given coordinate neighbourhood into the unit sphere S^2. This map is the <u>Gauss map</u> and as $\underline{x}$ ranges over the coordinate neighbourhood, $\underline{e}_3$ ranges over some region of S^2. In the case where $K(p) \neq 0$ at some point p of S it follows from the Inverse Function Theorem (using the fact that $K = \det(d\underline{e}_3)$) that $\underline{e}_3$ defines a coordinate neighbourhood on S^2 in a neighbourhood of the image $\underline{e}_3(p)$ of p under $\underline{e}_3$. Furthermore $\underline{e}_1, \underline{e}_2, \underline{e}_3$ may then be considered as a moving frame on S^2 with $\underline{e}_3$ normal to S^2 at each point. Then the equation

$$d\underline{e}_3 = \omega_3^1\underline{e}_1 + \omega_3^2\underline{e}_2$$

is the analogue for S^2 of the equation

$$d\underline{x} = \theta^1\underline{e}_1 + \theta^2\underline{e}_2.$$

The area form on S^2 is therefore

$$\omega_3^1 \wedge \omega_3^2 = \omega_1^3 \wedge \omega_2^3 = K\,\theta^1 \wedge \theta^2. \tag{4.2}$$

Thus if we denote the elements of area on S and S^2 by dA_S and dA_{S^2} respectively we have the well-known result

$$\frac{dA_{S^2}}{dA_S} = K. \tag{4.3}$$

Secondly, using (2.18) and the generalized Stokes' Theorem we can easily prove the Gauss-Bonnet Theorem for a triangle on S. Suppose that T is a triangle on S and suppose that T is contained in a coordinate neighbourhood. Then

$$\int_{\partial T} \kappa_g \, ds + \int_T K\theta^1 \wedge \theta^2 = \int_{\partial T} (d^2\phi + \omega_1^2) - \int_T d\omega_1^2 \text{ by (2.18) and (4.1)}$$

$$= \int_{\partial T} d\phi + \int_{\partial T} \omega_1^2 - \int_{\partial T} \omega_1^2 \text{ by Stokes' Theorem}$$

$$= \int_{\partial T} d\phi.$$

This latter integral is the measure of the change in ϕ around ∂T and if the internal angles of T are α, β, γ it is not hard to show that this is $\alpha + \beta + \gamma - \pi$, the angular defect. Thus we have the Gauss-Bonnet Theorem for a triangle

$$\int_{\partial T} \kappa_g \, ds + \int_T K dA_S = \alpha + \beta + \gamma - \pi. \tag{4.4}$$

In particular if T is a geodesic triangle so that $\kappa_g = 0$ on ∂T then we have the well-known result:

$$\int_T K dA_S = \alpha + \beta + \gamma - \pi. \tag{4.5}$$

Thirdly, from equation (4.1), we may readily verify the Theorema Egregium of Gauss which states that the Gaussian curvature K depends only on the First Fundamental Form. To see this let us first note that the 1-form ω_1^2 may be written as a linear combination of θ^1 and θ^2, say

$$\omega_1^2 = a_1\theta^1 + a_2\theta^2. \tag{4.6}$$

Then using the first two structure equations (2.7), (2.8) we have

$$d\theta^1 = a_1\theta^1 \wedge \theta^2, \quad d\theta^2 = a_2\theta^1 \wedge \theta^2 \tag{4.7}$$

so that a_1 and a_2 are determined by θ^1 and θ^2. But then ω_1^2 is entirely determined by θ^1 and θ^2 and hence in view of equation

(4.1) so is the Gaussian curvature K. Indeed we can use this argument to derive an explicit expression for K in terms of the First Fundamental Form. For simplicity let us suppose that we are given orthogonal coordinates so that F = 0. Then $ds^2 = E(du^1)^2 + G(du^2)^2$ and we let $\theta^1 = \sqrt{E}\, du^1$, $\theta^2 = \sqrt{G}\, du^2$.

(This corresponds to the choice of $\underline{e}_1 = \dfrac{\underline{x}_1}{\sqrt{E}}$, $\underline{e}_2 = \dfrac{\underline{x}_2}{\sqrt{G}}$.)

Then

$$d\theta^1 = -\frac{1}{2}\frac{E_2}{\sqrt{E}}\, du^1 \wedge du^2 = -\frac{1}{2}\frac{E_2}{\sqrt{E}} \cdot \frac{1}{\sqrt{EG}}\, \theta^1 \wedge \theta^2$$

$$d\theta^2 = \frac{1}{2}\frac{G_1}{\sqrt{G}}\, du^1 \wedge du^2 = \frac{1}{2}\frac{G_1}{\sqrt{G}}\frac{1}{\sqrt{EG}}\, \theta^1 \wedge \theta^2.$$

Thus $a_1 = -\dfrac{1}{2\sqrt{E}}\dfrac{E_2}{\sqrt{EG}}$, $a_2 = \dfrac{1}{2\sqrt{G}}\dfrac{G_1}{\sqrt{EG}}$ and so

$$\omega_1^2 = a_1\theta^1 + a_2\theta^2$$

$$= \frac{1}{2}\left\{ -\frac{E_2}{\sqrt{EG}}\, du^1 + \frac{G_1}{\sqrt{EG}}\, du^2 \right\}$$

and $$d\omega_1^2 = \frac{1}{2}\left\{\left(\frac{E_2}{\sqrt{EG}}\right)_2 + \left(\frac{G_1}{\sqrt{EG}}\right)_1\right\} du^1 \wedge du^2.$$

It follows now from (4.1) that

$$K = -\frac{1}{2\sqrt{EG}}\left\{\left(\frac{E_2}{\sqrt{EG}}\right)_2 + \left(\frac{G_1}{\sqrt{EG}}\right)_1\right\}. \tag{4.8}$$

The simplicity of the calculation here contrasts sharply with that of other methods. The classical argument along these lines involves expressing K in terms of the Christoffel symbols which are themselves shown to depend only on the metric. See [1]. This is, in essence, what we have done above but because of the use of both orthonormal frames and differential forms the calculation has been considerably simplified. We recall that the Christoffel symbols may be defined by

$$\underline{x}_{ij} = \Gamma^k_{ij}\underline{x}_k + L_{ij}\underline{N}. \tag{4.9}$$

By using differential forms the six Christoffel symbols (the fact that $\Gamma^k_{ij} = \Gamma^k_{ji}$ reduces the number of symbols from eight to six) are replaced by one symbol ω^2_1 or perhaps to be more accurate by the two coefficients a_1, a_2 in the expression $\omega^2_1 = a_1\theta^1 + a_2\theta^2$. This is not hard to see in terms of the covariant derivative since specifiying the covariant derivative for the surface is equivalent to specifying either the Christoffel symbols or the 1-form ω^2_1. In the former case the covariant derivative is determined once $\nabla_{\underline{x}_j}\underline{x}_i$, (that is the tangential component of $\underline{x}_{ij}$), is known for all values 1 and 2 of i and j which by (4.9) is equivalent to knowing the Christoffel symbols. In the latter case the covariant derivative is determined once $\nabla_{\underline{e}_j}\underline{e}_i$ is known for all values 1 and 2 of i and j and since $\nabla_{\underline{e}_j}\underline{e}_i = \omega^1_i(\underline{e}_j)\underline{e}_1 + \omega^2_i(\underline{e}_j)\underline{e}_2$ this is equivalent to knowing ω^2_1.

Next we consider the Codazzi-Mainardi equations (2.11), (2.12) and see how they compare with the classical formulation. Again for simplicity we consider the case of orthogonal coordinates and write $\underline{e}_1 = \dfrac{\underline{x}_1}{\sqrt{E}}$, $\underline{e}_2 = \dfrac{\underline{x}_2}{\sqrt{G}}$, $\theta^1 = \sqrt{E}\, du^1$, $\theta^2 = \sqrt{G}\, du^2$. Then taking the exterior derivative of the equations $\underline{x}_1 = \sqrt{E}\,\underline{e}_1$, $\underline{x}_2 = \sqrt{G}\,\underline{e}_2$ we have

$$\underline{x}_{11}du^1 + \underline{x}_{12}du^2 = \frac{1}{2}\frac{dE}{\sqrt{E}}\underline{e}_1 + \sqrt{E}\, d\underline{e}_1 \tag{4.10}$$

$$\underline{x}_{12}du^1 + \underline{x}_{22}du^2 = \frac{1}{2}\frac{dG}{\sqrt{G}}\underline{e}_2 + \sqrt{G}\, d\underline{e}_2. \tag{4.11}$$

Taking the dot product of each of these with $\underline{e}_3 = \underline{N}$ then gives:

$$Ldu^1 + Mdu^2 = \sqrt{E}\,\omega^3_1$$

$$Mdu^1 + Ndu^2 = \sqrt{G}\,\omega^3_2$$

so that

$$\omega_1^3 = \frac{1}{\sqrt{E}} \{ Ldu^1 + Mdu^2 \} \tag{4.12}$$

$$\omega_2^3 = \frac{1}{\sqrt{G}} \{ Mdu^1 + Ndu^2 \}. \tag{4.13}$$

Thus

$$d\omega_1^3 = \left\{ - \left(\frac{L}{\sqrt{E}} \right)_2 + \left(\frac{M}{\sqrt{E}} \right)_1 \right\} du^1 \wedge du^2 \tag{4.14}$$

$$d\omega_2^3 = \left\{ - \left(\frac{M}{\sqrt{G}} \right)_2 + \left(\frac{N}{\sqrt{G}} \right)_1 \right\} du^1 \wedge du^2. \tag{4.15}$$

Also from above

$$\omega_1^2 = \frac{1}{2} \left\{ - \frac{E_2}{\sqrt{EG}} du^1 + \frac{G_1}{\sqrt{EG}} du^2 \right\} \tag{4.16}$$

so that

$$\omega_2^3 \wedge \omega_1^2 = \frac{1}{2G\sqrt{E}} \left\{ MG_1 + NE_2 \right\} du^1 \wedge du^2 \tag{4.17}$$

$$\omega_1^3 \wedge \omega_2^1 = - \frac{1}{2E\sqrt{G}} \left\{ LG_1 + ME_2 \right\} du^1 \wedge du^2. \tag{4.18}$$

Thus from (4.14), (4.15), (4.17), (4.18) the Codazzi-Mainardi equations (2.11), (2.12) may be written as

$$- \left(\frac{L}{\sqrt{E}} \right)_2 + \left(\frac{M}{\sqrt{E}} \right)_1 + \frac{1}{2G\sqrt{E}} \left\{ MG_1 + NE_2 \right\} = 0$$

$$- \left(\frac{M}{\sqrt{G}} \right)_2 + \left(\frac{N}{\sqrt{G}} \right)_1 - \frac{1}{2E\sqrt{G}} \left\{ LG_1 + ME_2 \right\} = 0$$

which reduce to

$$L_2 - M_1 = \frac{1}{2} \left(\frac{L}{E} + \frac{N}{G} \right) E_2 - \frac{1}{2} M \left(\frac{E_1}{E} + \frac{G_1}{G} \right) \tag{4.19}$$

$$N_1 - M_2 = \frac{1}{2} \left(\frac{L}{E} + \frac{N}{G} \right) G_1 - \frac{1}{2} M \left(\frac{E_2}{E} + \frac{G_2}{G} \right) \tag{4.20}$$

which are the Codazzi-Mainardi equations in their standard form. Again observe that the expressions are more complicated than their equivalents in terms of differential forms and that it is not immediately clear what their significance is when expressed in this way.

REFERENCES

[1] do Carmo, M., *Differential Geometry of Curves and Surfaces*, Prentice Hall, 1976.

[2] Cartan, E., *Oeuvres complètes*, Gauthier-Villars, Paris, 1952-55. Reprinted by Springer, 1984.

[3] Flanders, H., *Differential Forms with Applications to the Physical Sciences*, Academic Press, 1963.

[4] O'Neill, B., *Elementary Differential Geometry*, Academic Press, 1966.

[5] Willmore, T.J., *An Introduction to Differential Geometry*, Oxford University Press, 1959.

SURFACE/SURFACE INTERSECTION PROBLEMS

M.J. Pratt
(Cranfield Institute of Technology)

A.D. Geisow
(Cambridge Interactive Systems Ltd)

1. INTRODUCTION

The calculation of intersection curves between general surfaces is one of the major current problems in computer aided design. There is a class of design systems called solid or geometric modellers, in which the object being designed is represented by means of a unified data structure in the computer (Requicha & Voelcker [30], Pratt [28]). This structure must contain details of all the faces, edges and vertices of the object. Faces lie in specified surfaces and are bounded by edges which lie on surface/surface intersection curves. These curves are easy to compute if all the surfaces are planes or cylinders, for example. However, solid modellers are increasingly acquiring the capability to represent objects with 'sculptured' or free-form surfaces, usually of the parametric types discussed elsewhere in these proceedings, and the intersection problem is more difficult when such surfaces are involved. Since solid modellers are usually interactive graphical systems, and since they are intended for practical use in an engineering environment, the implications are that surface intersection calculation should be

(i) accurate, in the usual numerical sense;

(ii) robust, in the sense that it is not subject to failure and that it finds all the portions of an intersection curve which may have many disjoint sections;

(iii) fast, because the user of an interactive system does not wish to sit for long periods awaiting the results of calculations.

Many methods have been suggested for computing intersection curves between surfaces, and some of them are reviewed in the remainder of this paper. A fair degree of reliability has been achieved, but the ideal combination of accuracy, robustness and speed has not yet been attained, and there is scope for much further research in this area.

There are several other problems closely related to the surface/surface intersection problem. These include contouring of surfaces, computation of profile or silhouette lines for the graphical rendering of surfaces and calculation of the path of a machine tool cutter moving in contact with two surfaces (Geisow [7], Faux & Pratt [6]). Some of the methods described in the ensuing sections can be modified to deal with these related problems.

2. GENERAL CONSIDERATIONS

2.1 Types of Surface Representation

It was noted earlier in these proceedings that curves and surfaces may be defined implicitly or parametrically. It is worth mentioning here that curves may be defined not only in terms of Euclidean coordinates but also in terms of the parametric coordinates of a specific parametric surface. The possible types of curve and surface representations we will consider are therefore as summarised in Table 2.1:

Table 2.1:
Curve and Surface Representations

TYPE	CURVE	SURFACE
IMPLICIT 2D	$f(x,y) = 0$	
3D	$F(x,y,z) = 0$, $G(x,y,z) = 0$	$f(x,y,z) = 0$
PARAMETRIC	$\underline{r} = \underline{r}(t)$	$\underline{r} = \underline{r}(u,v)$
CURVE ON PARAMETRIC SURFACE $\underline{r} = \underline{r}(u,v)$	$h(u,v) = 0$ or $\underline{r}(t) = \underline{r}(u(t),v(t))$	

Implicit and parametric representations both have advantages and disadvantages when used for computer aided design. These may be illustrated by comparing two frequently occurring types of geometric calculation:

(i) Find a point on a surface corresponding to specified values of the independent variables. If the surface is parametrically represented as $\underline{r} = \underline{r}(u,v)$ it is only necessary to substitute values of u and v, then evaluate the right-hand side. An implicit surface equation $f(x,y,z) = 0$ cannot always be solved explicitly for z as a function of x and y (assuming these to be the independent variables). It is often necessary to solve a nonlinear equation for z once x and y have been specified. Admittedly this is only of quadratic degree for the commonly occurring quadric surfaces, but even then there is the problem of choosing the appropriate sign for the square root of the discriminant in the solution. The parametric form is clearly the more convenient when surface points are to be computed.

(ii) Determine whether a given point X,Y,Z lies on a specified surface. When the surface is implicitly defined as $f(x,y,z) = 0$ it is only necessary to substitute X,Y and Z to determine whether or not they satisfy the equation. In the parametric case, however, all we can do easily is compute points which do lie on the surface. A search procedure must be used to determine whether X,Y,Z is a point on the surface; it will also give the nearest point if X,Y,Z lies off the surface, but this is clearly a much less convenient kind of calculation than for the implicitly defined surface.

2.2 *Influence of Surface Representation on the Intersection Problem*

Three cases of the intersection problem arise, distinguished by the types of representation of the two surfaces involved. In order of increasing difficulty the cases are as follows:

(i) implicit/parametric.

If the surface equations are

$$f(x,y,z) = 0$$

and

$$\underline{r} = \underline{r}(u,v) = [x(u,v), y(u,v), z(u,v)]$$

then simple substitution gives

$$f(x(u,v), y(u,v), z(u,v)) = F(u,v) = 0.$$

This is the simplest case. The curve is represented by a single equation in two variables, and it is defined in the parameter space of the second surface.

(ii) implict/implicit.

The surface equations are

$$f(x,y,z) = 0$$

and

$$g(x,y,z) = 0,$$

where f and g are usually nonlinear. In this case the curve is defined in the classical manner by two equations in three variables.

(iii) parametric/parametric.

The surface equations are

$$\underline{r} = \underline{r}_1(u_1,v_1)$$
$$\underline{r} = \underline{r}_2(u_2,v_2).$$

For points lying on both surfaces we have

$$\underline{r}_1(u_1,v_1) - \underline{r}_2(u_2,v_2) = \underline{0} ,$$

and the curve is therefore represented by three scalar equations in the four variables u_1,v_1,u_2,v_2. Again, the equations are usually nonlinear.

In all three cases it is usual to compute points on the intersection curves by specifying the value of one variable, or imposing some more general additional constraint, and solving the resulting determinate system of equations numerically. This matter is discussed further in Section 3.3.

It has been shown that the mathematical nature of the intersection problem is simplest when one surface is represented implicitly, the other parametrically. This raises the question as to whether it is possible to convert one of the surfaces from implicit to parametric form or vice versa in thoses cases where both surfaces are of the same type. This would give a mathematical problem of class (i), the simplest of the classes listed above. Alternatively, it may be possible to reduce a class (ii) problem to one of class (i) by algebraic elimination of one of the variables, and similarly to simplify a class (iii) problem by elimination of two of the variables. These possibilities appear to be feasible in

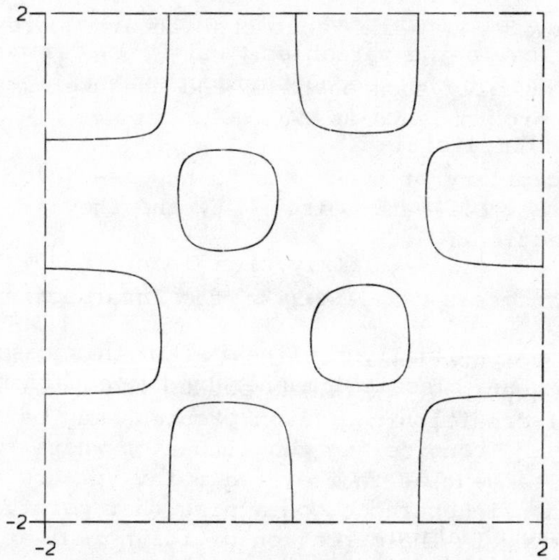

Fig. 2.1: Plane (z=0) Contours of the Bicubic Surface $z = (x-1)x(x+1)(y-1)y(y+1) + 0.05$ on the Rectangle $[-2,2] \times [-2,2]$.

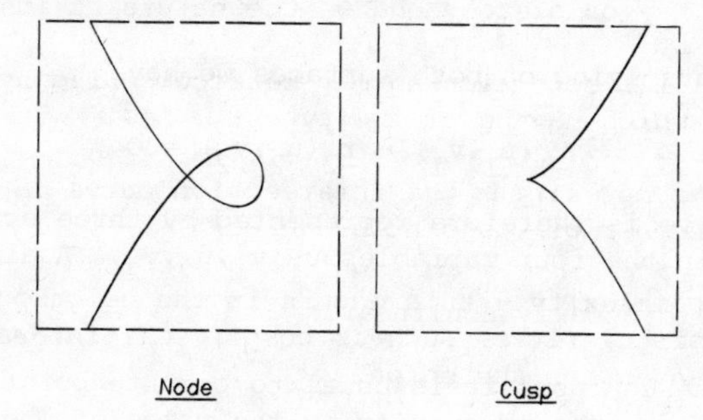

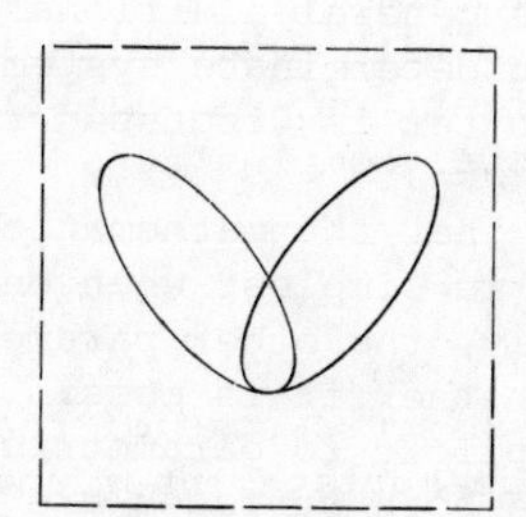

Fig. 2.2: Cusped, Self-intersecting and Self-tangent Curves: some examples of Local Complexity

principle provided all the functions concerned are polynomials or rational polynomials, but the algebraic complexity of the reduction is usually such as to demand the use of a symbolic manipulation program such as MACSYMA.

The applicability of such methods has been demonstrated by Sederberg et al. [39] and Geisow [7], and they are examined further in Section 3.1.

2.3 Conditioning and Complexity of the Intersection Problem

Since the computation of intersection curves involves the solution of equations, it is subject to the usual problems of numerical ill-conditioning. Such problems may be illustrated geometrically by considering the situation where two surfaces are mutually tangent at an isolated point but are almost parallel in a neighbourhood containing that point. If one surface is moved in the direction of its normal at the point of tangency then either there will be no point common to both surfaces or the original point will become a curve of intersection. Thus small errors in the numerical treatment of the problem may completely change the nature of its solution.

Numerical ill-conditioning may be accompanied by geometric complexity, which may be of two types:

(i) global complexity - the intersection curve may have a large number of disjoint branches;

(ii) local complexity - this occurs in the neighbourhood of curve singularities such as cusps, self-intersections, self-tangencies and so on.

An example of global complexity is provided by the intersection of a plane with a bicubic algebraic surface, which can have up to eight branches, as illustrated in Fig. 2.1. Some examples of local complexity are shown in Fig. 2.2.

3. TECHNIQUES

In this section we discuss some of the techniques which have been applied to surface/surface intersection problems. Geisow [7] distinguishes four main categories of method, namely algebraic, lattice evaluation, marching and recursive subdivision. They are dealt with under separate subheadings.

3.1 Algebraic Methods

For certain simple cases surface intersection curves may be available explicitly. Plane/plane and plane/quadric intersections, for example, give rise to straight lines and conics respectively, and these may be calculated directly in whatever form is most useful. In more complex situations some preliminary algebraic manipulation may help to put the problem in a more convenient form for computation, even though an explicit solution is not available.

If one surface is parametric and the other implicit, then their intersection curve can be represented implicitly in the parameter space of the first surface by simple substitution, as mentioned earlier in Section 2.2.

For the case of two implicit surfaces elimination of one unknown, say z, between their two equations gives a result of the form $h(x,y) = 0$, which is the projection of the intersection curve onto the x,y-plane. An alternative approach would be to convert one surface to parametric form and then use substitution. However, the most complex parametric forms we wish to deal with in practice are of rational polynomial type, and unfortunately not all algebraic implicit surfaces can be parametrised in this way. All quadric surfaces do possess a rational parametric form, and this fact was used by Levin [15], together with an additional simplification. If two implicit surfaces $f = 0$ and $g = 0$ intersect in a curve C then $\alpha f + \beta g = 0$, where α,β are any real scalars, will also contain C. It is always possible to choose α and β so that the linear combination is a ruled quadric. Levin substituted a canonical representation of this surface into one of the original quadrics to obtain a description of the intersection curve which was quadratic in one of the variables.

As shown in Section 2.2 the intersection curve between two parametric surfaces $\underline{r} = \underline{r}_1(u_1,v_1)$ and $\underline{r} = \underline{r}_2(u_2,v_2)$ is

$$\underline{r}_1(u_1,v_1) - \underline{r}_2(u_2,v_2) = \underline{0}.$$

This represents three scalar equations in four unknowns. Two of the unknowns, say u_2,v_2, can in principle be eliminated to leave a single implicit equation $h(u_1,v_1) = 0$, describing the intersection curve in the parameter space of one of the surfaces. Sederberg [39] has recently suggested an approach based on conversion of one of the parametric surfaces to implicit form by elimination of the two surface parameters between the three component equations of the surface. The relative merits of this approach against the direct elimination approach have not apparently been examined, but in either case the algebraic complexity is enormous for the commonly occurring parametric bicubic surfaces. A bicubic patch proves to be equivalent to an algebraic surface of degree 18, whose equation contains 1330 terms. The intersection curve between two such surfaces has geometric degree 324, i.e. it can have as many as 324 intersections with a plane.

Elimination of an unknown between two multivariate polynomial equations can be achieved by forming their 'resultant', which may be expressed in various forms using theoretical results established in the last century. Given two polynomial equations

$$f(x) = a_0 x^n + a_1 x^{n-1} + \text{----} + a_n = 0$$

and $$g(x) = b_0 x^m + b_1 x^{m-1} + \text{----} + b_m = 0,$$

where a_0, b_0 are non-zero and the coefficients may be functions of other variables, we wish to obtain an equation not containing x which holds if and only if both original equations hold. Suppose that the zeros of f(x) are $\alpha_1, \alpha_2, \text{---}, \alpha_n$. If one of the α_i is also a zero of g(x) then

$$g(\alpha_1) g(\alpha_2) \text{----} g(\alpha_n) = 0,$$

and the vanishing of the expression on the left-hand side is necessary and sufficient for f(x) and g(x) to have a common zero. The resultant of the two original equations is defined to be a multiple of this expression:

$$R(f,g) = a_0^m g(\alpha_1) g(\alpha_2) \text{---} g(\alpha_n),$$

and R(f,g) = 0 is an equation from which x is absent but which holds whenever both original equations are satisfied.

A convenient determinantal form for R(f,g) was given by Sylvester:

$$R(f,g) = \begin{vmatrix} a_o & a_1 & a_2 & \cdot & \cdot & \cdot & a_n & & & \\ & a_o & a_1 & \cdot & \cdot & \cdot & \cdot & a_n & & \\ & & \cdot & \cdot & \cdot & \cdot & \cdot & \cdot & \cdot & \\ & & & a_o & a_1 & \cdot & \cdot & \cdot & \cdot & a_n \\ b_o & b_1 & b_2 & \cdot & \cdot & \cdot & b_m & & & \\ & b_o & b_1 & \cdot & \cdot & \cdot & \cdot & b_m & & \\ & & \cdot & \cdot & \cdot & \cdot & \cdot & \cdot & \cdot & \\ & & & b_o & b_1 & \cdot & \cdot & \cdot & \cdot & b_m \end{vmatrix} \quad \begin{matrix} \text{(m rows)} \\ \\ \text{(n rows)} \end{matrix}$$

It is clear from this form that $R(f,g)$ and $R(g,f)$ have the same magnitude though their signs may differ; the order of the arguments is therefore immaterial in practice. Further details and proofs can be found in Uspensky [41] or in older texts such as Salmon [37]. Other expressions for the resultant have also been derived. In particular, Sederberg has used a form due to Cayley and Bezout. Although these elimination methods have not received wide attention in the field of computational geometry they are by no means unknown; Weiss [43] treated the quadric/quadric intersection problem by eliminating z between the implicit forms $f(x,y,z) = 0$ and $g(x,y,z) = 0$, using Sylvester's determinant to calculate the resultant of f and g. This gave a quartic equation $h(x,y) = 0$; successive values of x were then substituted into h and solutions for y determined. These (x,y) pairs were finally substituted back into f and g to find common solutions for z.

Woon [44] determined planar intersection curves between quadrics by elimination (again using Sylvester's determinant); the plane was computed by finding an appropriate linear combination of the original surfaces and splitting this into the product of two linear factors. One of these could then be used in the elimination, thus yielding a conic rather than a quartic curve.

Sabin [35], also dealing with quadric surfaces, calculated the projection of the intersection curve onto a viewing plane by looking for common intersections of a line of sight with both surfaces. For a parallel projection, this is equivalent to eliminating z by taking the resultant of the two surfaces (q.v. Weiss). Sabin also suggests the extension of his method to higher order implicit surfaces. He does point out a general disadvantage of eliminating one of the space variables; a certain amount of information has been lost by the projection about the configuration of the curve in space. This information must be regained by substituting points on the projected curve back into the surfaces and looking for equal roots, which involves quite a lot of extra work.

All the above mentioned approaches produce some function of two variables describing the solution curve either as a projection or embedded in a parametric surface. The generation of points on such curves usually presents some difficulty; lattice evaluation as described in the next Section was used in most of the above cases, and to quote Sabin [35] "Computational difficulties can occur when a curve appears to cross itself. If a coarse lattice is used, the curve may be drawn with incorrect connectivity."

3.2 Lattice Evaluation

This is a technique (also known as discrete evaluation or space grid) based on an initial reduction of the number of degrees of freedom in a problem. A set of discrete values is substituted for some of the variables to obtain a number of sub-problems of lower dimension. The solutions to these must then be juxtaposed and interpolated to give an approximation to the required overall solution. Several contouring algorithms (e.g. Heap [10], Payne [22,23]), work by evaluating a matrix of values $z_{ij} = f(x_i, y_i)$ and tracing a particular $z = h$ contour through the corresponding grid (see Fig. 3.1). If the z values at the two ends of a cell edge bracket h then the desired contour crosses that edge. The crossing point is found by inverse interpolation. The contour entering the cell across this edge must leave by crossing one of the other edges; the second crossing point is similarly located and the contour traced through the cell to connect the two points. The process then steps to the neighbouring cell across the second edge. Some problems can arise if the contour crosses all four edges, when the incorrect connectivity may be chosen. Contours which enter and leave a cell by the same edge will not be correctly traced, as the z-values at the ends of that edge will not bracket the contour height. Small loops within cells will be completely missed.

South and Kelly [40] used lattice evaluation to contour bicubic patches; they substituted the parametric patch into an implicit representation of the plane to obtain a bicubic implicit equation in the parameter domain, and then traced the curve in the manner described above.

McRae [17] devised two algorithms for sectioning biquadratic Bezier patches. The first, which he rejected because of lack of robustness, used an adaptive semi-discrete approach. An intersection point was found on one edge of a patch and the curve then traced across by substituting discrete values for one of the parameters, finding the intersection points of the corresponding constant parameter curves with the plane and finally joining them in a piecewise linear manner. Difficulties were encountered if the intersection curve was tangential to a patch edge, or if it was entirely interior to the patch, in which case it was not found at all. Adjustable tolerances were provided, but the algorithm was found difficult to 'tune' for optimal performance. McRae's more satisfactory second algorithm is based on recursive subdivision, and is described in Section 3.4.

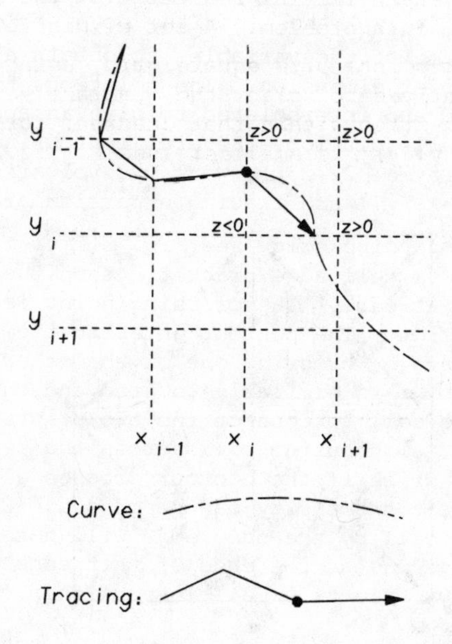

Fig. 3.1 Lattice Evaluation Method

The problems with lattice methods may be summarised as follows:

(i) It can be difficult to choose the correct connectivity between discrete solutions, especially near saddle points and other singularities.

(ii) Features may get lost in the holes in the lattice - a loop will be missed unless there is at least one grid point evaluated inside the loop. Using a finer grid increases the computation time, and even then does not guarantee detection of these 'lost' details.

3.3 Marching Methods

Marching techniques generate a sequence of points on or near the required intersection curve by stepping from the current point in a direction controlled by the local differential geometry of the surface or surfaces involved. The simplest methods of this class are those sometimes used for displaying curves on raster displays or incremental plotters, where the points generated all have integer coordinates and form a dense chain. Jordan, Lennon & Holm [11] developed an algorithm of this kind for directly displaying a curve $f(x,y) = 0$. From a given start point (x_o, y_o) the algorithm steps to one of the neighbouring eight points of a square grid, using the signs of the partial derivatives to select the quadrant moved to, and stepping to the position within that quadrant for which the value of $|f(x+\delta x, y+\delta y)|$ is smallest (see Fig. 3.2.).

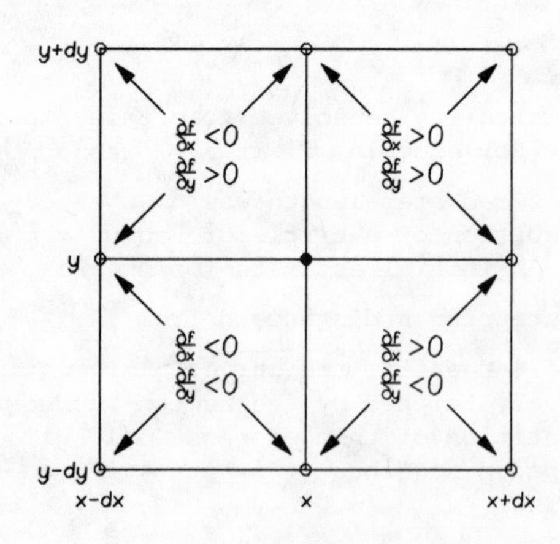

Fig. 3.2 Discrete Marching

These methods work well for straight lines and conics, but run into problems with more complex curves. Finding suitable starting points and termination conditions becomes harder, and the algorithms usually get stuck if they encounter stationary points of the surface, for at these points the first order partial derivatives vanish.

A second class of marching methods may be thought of as repeatedly solving a number of simultaneous equations, one of which is a 'step constraint'. The intersections we are considering have two degrees of freedom, one of which is removed by imposing an extra arbitrary condition on the system. This is usually chosen to control the spacing of the solution points in a suitable manner. In the case of surface-surface intersections, the extra condition may be thought of as a third surface which intersects the solution curve at a distance from the 'current' point. A similar concept is used in numerically controlled machining using programs such as APT (Faux and Pratt [6]). Here the cutting tool is driven in contact with a part surface and a drive surface which correspond to the two intersecting surfaces, while a sequence of 'pseudo-check surfaces' is used to provide step contraints.

If the intersection problem can be reduced to the implicit form

$$f(x,y) = 0,$$

then the addition of a step constraint

$$g(x,y) = 0$$

gives a system which may be solved to obtain a point on the solution curve. For example, if (x_o,y_o) was a point satisfying $f = 0$ and the desired step-length was d, then $g = 0$ could be chosen as the equation of a circle of radius d centred on the point (x_o,y_o). An initial estimate for the next point could be obtained by stepping a distance d from (x_o,y_o) in the direction of the tangent to the curve. This estimate could then be iteratively refined by the Newton-Raphson method (see Fig. 3.3). Repetition of this process would give an equidistantly spaced sequence of points lying on the intersection curve.

If the intersection curve is the intersection of two implicit surfaces then we may proceed in a similar manner, but with a step constraint of the form

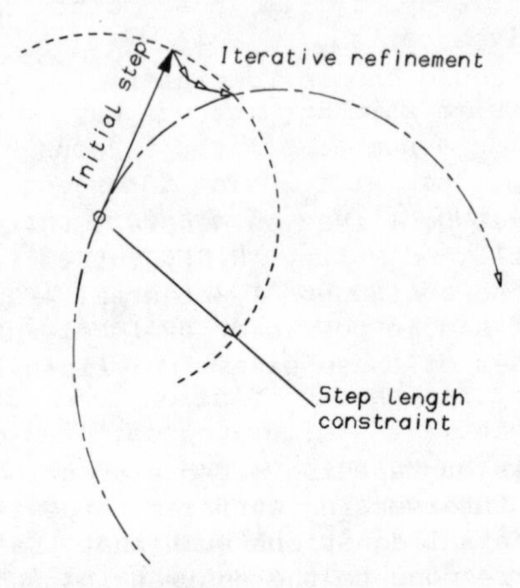

Fig. 3.3 **Marching Method**

for example a sphere of radius 'd' centred on the point (x_o, y_o, z_o). This time the Newton-Raphson iteration is in three dimensions. For the intersection of two parametric surfaces we have three equations in four unknowns; and, in general, the step constraint will be some function of these four parameters.

The example step constraint conditions given here are fairly primitive; for practical purposes, a smoother looking curve could be obtained more economically by making the step length vary with the curvature of the solution curve (see Sabin [32], for example).

Woon [44] traced the intersection curves of implicitly defined quadrics using a simplified marching technique, stepping by a fixed increment in one of the variables x, y or z.

A third class of marching methods is obtained by minimising some function L which is zero on the intersection curve and positive elsewhere. For surface and step constraint equations

$$e(x,y,z) = 0$$

$$f(x,y,z) = 0$$

$$g(x,y,z) = 0$$

for example, we could choose to minimise

$$L = e^2 + f^2 + g^2.$$

Much work has been published on general minimisation methods (Powell [26], Gill and Murray [8,9], Butterfield [3]), mainly concerned with the avoidance of potential problems, notably the detection of saddle-points or of local rather than global minima. In either of these cases the original equations are not satisfied.

A fourth class of marching methods uses local differential geometry of the intersecting surfaces to construct a system of ordinary differential equations such that the solution trajectories correspond to the desired intersection curves. Phillips and Odell [25] use a standard ODE package to solve a system of three equations in computing the intersection of two implicitly defined surfaces. The method used to compute cutter paths in the APT program for the numerical control of machine tools in effect uses a similar method, since it relies on the fact that the direction of the intersection curve at any point is always perpendicular to the plane containing both surface normals at that point (Faux and Pratt [6]).

The following problems are common to all the marching methods discussed:

(i) A start point must be found before stepping can begin. Start points are usually sought along the boundary of the region of interest, but internal loops will then be missed. A grid based search can be used, similar to that of the lattice methods, but there is still a danger of missing small loops, and care must be taken not to trace the same branch of a curve more than once by starting at different points along the same curve.

(ii) The criteria for stepping must be chosen carefully. A curve may form a closed loop, and the method should stop if it comes back close to its start point. Similarly, the curve may leave the region of interest, which for intersections of parametric surfaces would normally be $[0,1]\times[0,1]$. To avoid these problems, checks must be made after each step has been taken to compare the new point with the boundary of the region and with the first point of the current branch. If the curve is self-intersecting, the stepping process may choose the 'wrong' exit from the self-intersection point. This can cause looping.

(iii) Too large a step length allows the possibility of stepping to a neighbouring branch of the curve, which gives the wrong connectivity (see Fig. 3.4a). Worse, the next step may then be back towards the previous point, causing endless looping (Fig 3.4b).

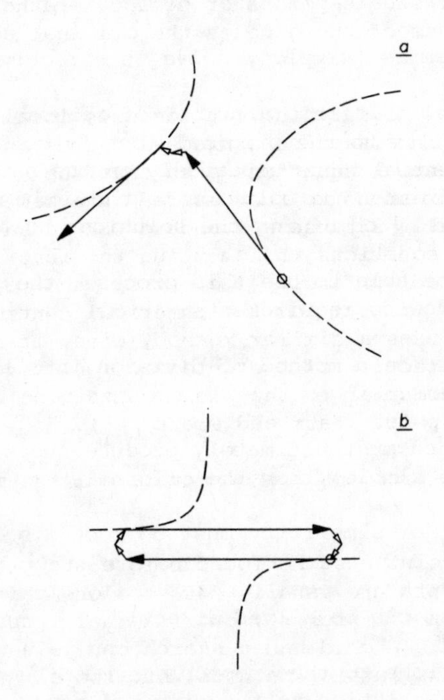

Fig. 3.4 Stepping Failures

These difficulties have all been tackled to some extent (Sabin [31, 32, 33, 34]) and algorithms based on marching can perform very well when correctly 'tuned'. However, a satisfactory compromise between robustness and efficiency still remains an elusive goal.

3.4 Recursive Subdivision

Recursive subdivision techniques are based on the 'divide and conquer' paradigm, and their use in solving intersection problems is most appropriate with the Bezier and B-spline curve and surface formulations. The use of such methods in computational geometry goes back to the hidden line algorithm of Warnock [42], since when they have steadily gained in popularity. Catmull [4], and Lane, Carpenter, Whitted and Blinn [12], have used subdivision for displaying shaded pictures of parametric surfaces. Lane and Riesenfeld [13] first suggested the application of subdivision techniques to intersection problems. The basic idea may be expressed as follows:

If the problem is 'simple', solve it directly.

Otherwise, split it into a number of subproblems, each of which is similar to the original (but 'simpler' in some sense) and tackle these separately in the same way as the original problem. The solution to the original problem is then obtained by combining the solutions to the subproblems.

In order to be able to use this process, the problem must satisfy the following requirements:

(i) There must be a method of division into like subproblems.

(ii) This method must ultimately produce subproblems which are 'more simple' than the original.

(iii) The level of simplicity must be recognisable.

(iv) There must be a level of simplicity at which the subproblem can be solved directly.

(v) The solutions to the subproblems must be combinable to form a solution to the original problem.

The Bezier and B-spline formulations define parametric curves and surfaces in terms of control polygons and control polyhedra respectively (Pratt [29]). Recursive subdivision in general is dealt with elsewhere in these Proceedings (Sabin [36]), but we illustrate here for the simple specific case of a quadratic Bezier curve (see Fig. 3.5). This curve may be split into two pieces at a parameter value t = r, and each piece defined in terms of a new smaller, control polygon whose points lie closer to the curve (Fig. 3.5b). The curve between t = 0 and t = 1 lies inside the convex hull of the control polygon; this is evident from the subdivision construction, which forms a sequence of convex combinations of the control points for r in [0,1]. The convex hull can be used as a coarse model of the curve; it becomes progressively more accurate as the curve is subdivided. Further, the curve has the variation diminishing property, which means that it will intersect a straight line no more times than the polygon does. In particular, if the polygon intersects a straight line exactly once, then so will the curve.

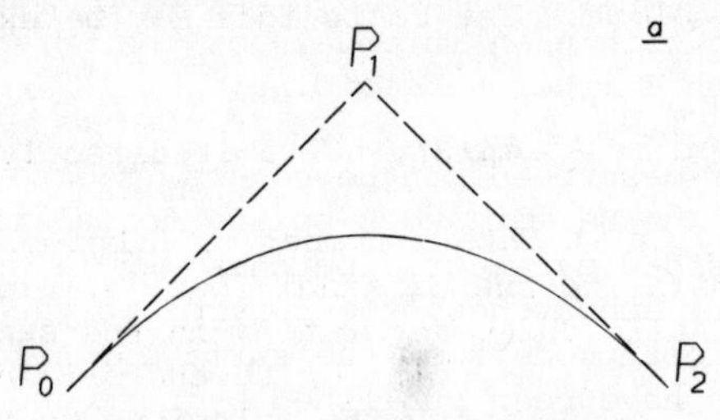

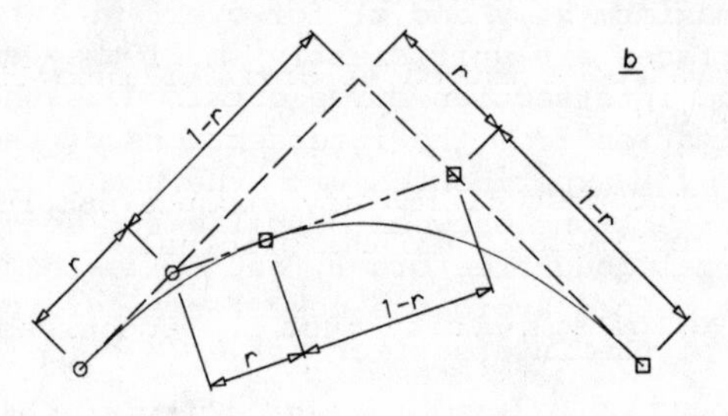

Fig. 3.5 Subdivision of Bezier curve

These considerations lead to a recursive subdivision algorithm for locating the intersection points of a straight line and a Bezier curve:

1. If the original control polygon does not cross the line, there is no intersection.

2. If the control polygon crosses the line once, there is one intersection, which lies within a known interval.

3. If the polygon crosses the line more than once, split the curve into two pieces, and examine the control polygons of each piece in turn using the same algorithm.

When an interval has been found which contains just one intersection point, that point may be located more accurately either by further subdivision or by other numerical techniques, such as the Newton-Raphson method. Repeated roots will cause the algorithm as described to recurse endlessly, as they cannot be separated. The usual solution to this is to put a lower limit on the size of an interval, below which it is regarded as being a point. We have now introduced two definitions of 'simplicity' for a subproblem - either there is only one root within an interval, or the interval is too small to be worth further subdividing.

The technique described can be generalised to surface intersections. The Bezier and B-spline formulations of bivariate four-sided parametric patches may be subdivided to yield two or four subpatches, the control polyhedra of which may be used as local models of the surface. Algorithms based on this approach have been published by Lane and Riesenfield [13], Cohen, Lyche and Riesenfield [5], and Boehm [1,2]. For surface/surface intersection, the usual approach has been to subdivide the surfaces and compare the bounding boxes (minimum and maximum x, y and z) for overlap. When the subdivided surfaces are sufficiently small they are considered planar, and the intersection curve obtained as the piecewise linear approximation from the intersections of these flat facets. Peng [24] experimented with the use of subdivision to locate intersections between B-spline surfaces, subdividing not in the normal quad-tree order, but following the curve once located. This helped avoid the need for a sorting phase to join portions of the intersection curve.

McRae [17] dealt with the sectioning of a biquadratic Bezier patch by comparing the control points of the patch with the plane equation. The control points were substituted into the plane equation, and the signs of the resulting values used to determine the nature of the intersection.

Geisow [7] proposed a general scheme for any interrogation which could be reduced to some polynomial form $I(u,v) = 0$, and was not restricted to Bezier or B-spline surfaces. The 'interrogation function' $I(u,v)$ may be expressed in terms of a bivariate Bernstein basis; recursive subdivision can then be used to isolate all 'simple' portions of the intersection curve, within which other methods (e.g. marching) could safely be used.

Lee and Fredricks [14] also use a subdivision technique which is not restricted to use with Bezier or B-spline surfaces. Their algorithm computes intersections between a general patched parametric surface and a plane. The patches are first subdivided and subpatches isolated which contain 'simple' curve segments. Within each subpatch the curve segment is approximated by a parametric cubic expressed in terms of the parametric u,v coordinates of the surface. The local differential geometry of the surface is used during this phase. If the approximation is unacceptable the curve segment is subdivided as many times as necessary to meet required tolerances. This algorithm has the interesting feature that a continuous representation of the curve is generated, once computed segments are correctly joined together. Such a representation can usually be stored much more economically than a lengthy sequence of computed points.

Experience has shown that the recursive subdivision approach is potentially very robust and reasonably efficient, provided the simplicity conditions can be reliably and cheaply recognised. The choice of simplicity conditions is of paramount importance; unless the chosen criteria can cope with all possible types of problem complexity, methods based on them will not be reliable. Unfortunately, not all published algorithms are ideal in this respect.

3.5 Homotopy Continuation

One technique has been used for the calculation of intersections which does not fall clearly under any of the previous four headings. The method is called homotopy continuation, and its use in the present context has been described by Morgan [18,19]. It may be used to solve systems of simultaneous polynomial equations, and conditions can be imposed which ensure that all the solutions are found.

The principle is to start with a simple problem with known solutions, and to modify this problem incrementally until the original more complex problem results, meanwhile following the behaviour of the solutions as the modification proceeds. Each solution traces out what is known as a homotopy path, which can be expressed as a solution curve of a certain ordinary differential equation related to the underlying algebraic equation.

The following example is given by Morgan [18]. The equations

$$x^2 - xy - 1 = 0$$

and

$$y^2 - x - 5 = 0$$

are originally simplified to give the uncoupled equations

$$x^2 - 1 = 0$$

and
$$y^2 - 5 = 0,$$

with obvious solutions $x = \pm 1$, $y = \pm \sqrt{5}$. The solutions are now computed, in the manner described, to the problem

$$x^2 - t\,x\,y - 1 = 0,$$

$$y^2 + t\,x - 5 = 0,$$

for a sequence of values of t which increase in small uniform steps from 0 to 1. The computed solution values give points on the homotopy paths of the four solutions as t increases.

The homotopy continuation method is suitable for computing intersections between implicitly defined surfaces if some form of step constraint is used to give a determinate set of three equations. It is reportedly used in the solid modeller GMSOLID developed by General Motors for the calculation of intersection points between three quadric surfaces. In this system intersection curves between two quadric surfaces are computed using a modification of Levin's method as described in Section 3.1 (Sarraga [38]).

4. REPRESENTATION OF THE COMPUTED INTERSECTION CURVES

Most of the methods discussed earlier compute a sequence of points on the desired intersection curve, which immediately defines a piecewise linear approximate representation of the curve. This is a practical form for some purposes, although it suffers from the following disadvantages:

(i) it requires the storage of a large volume of data,

(ii) the form is inconvenient if the computed curve is to be used for subsequent geometric construction, e.g. as one boundary of a new surface, and

(iii) important information can be lost. For example, two intersection curves may just fail to intersect while their polygonal approximations do intersect; conversely, it is possible for the true curves to intersect while their approximate representations do not.

The first two disadvantages may be avoided by the use of data reduction algorithms which construct piecewise polynomial curves with degree > 1 in terms of a much smaller amount of data. The use of standard least-squares methods in this context is complicated by the fact that the piecewise representations computed must usually be parametric in nature (Pratt [27]). Further, it is often necessary to impose tolerances not only on the maximum deviation of the curves from the computed data but also on deviations of tangent and curvature estimates (Nilsson [20]). Some work has been published in this area (Opheim [21], Lozover & Preiss [16]), but there is scope for further research. In general, different applications will require different data reduction algorithms. It should be noted that the algorithm of Lee and Fredricks mentioned earlier is unusual in generating a continuous curve representation from the outset. This may be a pointer to the future.

Where intersections of parametric surfaces are concerned there is another aspect to the curve representation problem. Points on the computed curve may be generated either in Euclidean 3D space or in the 2D parameter space of one (or both, if the second surface is not implicit) of the surfaces involved. In the first case the derived piecewise linear representation, or any economised version obtained by data reduction, will not lie on either of the intersecting surfaces. If the curve representation is based on parametric coordinates then the approximation to the true curve is constrained to lie on one or other of the surfaces, but not of course both. The best choice for practical purposes again depends on the subsequent application of the computed curve. Any system designed for engineering purposes should probably be capable of generating and storing any of the three possible forms.

5. CONCLUSION

A variety of methods for computing intersections between surfaces has been reviewed in this paper. It has been shown that several types of complexity bedevil the problem in general. The basic approaches have been classified into four main categories; many practical methods use a combination of strategies, however. None of the methods so far devised seems to achieve the ideal combination of accuracy, robustness and speed, although in many practical systems for computer aided design algorithms have been implemented which perform reasonably well most of the time. There is great scope for additional work in this general area, not only on the computation of the intersection curves themselves but on the development of suitable economised representations of them for practical purposes.

REFERENCES

[1] Boehm, W. Inserting New Knots in B-spline Curves, Computer Aided Design, **12**, pp. 199-201, 1980.

[2] Boehm, W., Generating the Bezier Points of B-spline Curves and Surfaces, Computer Aided Design, **13**, pp. 365-366, 1981.

[3] Butterfield, K.R., The Development and Application of Algorithms associated with Surface Representation, PhD Thesis, Brunel University, 1978.

[4] Catmull, E.A., A Subdivision Algorithm for Computer Display of Curved Surfaces, Report UTEC-CSc-74-133, University of Utah, 1978.

[5] Cohen, E., Lyche, T. & Riesenfeld, R., Discrete B-splines and Subdivision Techniques in Computer Aided Geometric Design and Computer Graphics, Computer Graphics and Image Processing, **14** pp. 87-111, 1980.

[6] Faux, I.D & Pratt, M.J., *Computational Geometry for Design and Manufacture,* Ellis Horwood, Chichester, 1979.

[7] Geisow, A.D., Surface Interrogations, PhD Thesis, University of East Anglia, School of Computing Studies and Accountancy, 1983.

[8] Gill, P.E. & Murray, W., Safeguarded Steplength Algorithms for Optimization using Descent Methods, Report NAC 37, National Physical Laboratory, Teddington, Middlesex, 1974.

[9] Gill, P.E. & Murray, W., Algorithms for the Solution of the Nonlinear Least Squares Problem, Report NAC 71, National Physical Laboratory, Teddington, Middlesex, 1976.

[10] Heap, B.R., Two FORTRAN Contouring Routines, Report NAC-47, National Physical Laboratory, Teddington, Middlesex, 1974.

[11] Jordan, B.W., Lennon, W.J. & Holm, B.D., An Improved Algorithm for the generation of Non-parametric Curves,

IEEE Transactions on Computers, **C-22**, pp. 1052-1060, 1973.

[12] Lane, J.M., Carpenter, L.C., Whitted, T & Blinn, J.F., Scan Line Methods for displaying Parametrically Defined Surfaces, Comm. ACM **23**, pp. 23-34, 1980.

[13] Lane, J.M. & Riesenfeld, R.F., A Theoretical Development for the Computer Display and Generation of Piecewise Polynomial Surfaces, IEEE Trans. Pattern Analysis & Machine Intelligence **2**, pp. 35-46, 1981.

[14] Lee, R.B. & Fredricks, D.A., Intersection of Parametric Surfaces and a Plane, IEEE Computer Graphics and Applications, **4**, 8, pp. 48-51, 1984.

[15] Levin, J.Z., A Parametric Algorithm for drawing pictures of Solid Objects composed of Quadric Surfaces, Comm. ACM **19**, pp. 555-563, 1976.

[16] Lozover, O. & Preiss, K., Automatic Generation of a Cubic B-spline Representation of a Digitised Curve, in *Proc. Eurographics 81 Conf.* (ed. J.L. Encarnacao), North-Holland Publ. Co., 1983.

[17] McRae, A.P., A Robust Algorithm for Sectioning Biquadratic Patches, Report CGP 79/4, Computational Geometry Project, University of East Anglia, 1979.

[18] Morgan, A.P., A Method for Computing All Solutions to Systems of Polynomial Equations, Report GMR-3651, General Motors Research Laboratories, Warren, Michigan, 1981.

[19] Morgan, A.P., A Method for Computing all Solutions to Systems of Polynomial Equations, ACM Trans. Mathematical Software, **9**, pp. 1-17, 1983.

[20] Nilsson, B., Preliminary Algorithm for Reduction of the Number of Points in FORMELA Curves in Saab-Scania's IDS-routine, Internal Document FMKG-81.107, Saab-Scania Aerospace Division, Linkoping, Sweden, 1981.

[21] Opheim, H., Smoothing a Digitised Curve by Data Reduction Methods, in *Proc. Eurographics 81 Conf.* (ed. J.L. Encarnacao), North-Holland Publ. Co., 1983.

[22] Payne, P.J., A Contouring Program for Coons Surface Patches, CAD Group Document No. 16, Cambridge University Engineering Dept., 1968.

[23] Payne, J.P., A Contouring Algorithm for Joined Surface Patches, CAD Group Document No. 58, Cambridge University Engineering Dept., 1971.

[24] Peng, Q.S., An Algorithm for finding the Intersection Line between two B-spline Surfaces, Computer Aided Design, **16**, 4, 191-196, 1984.

[25] Phillips, M.B. & Odell, G.M., An Algorithm for Locating and Displaying the Intersection of Two Arbitrary Surfaces, IEEE Computer Graphics and Applications, **4**, 9, pp. 48-58, 1984.

[26] Powell, M.J.D., Problems Related to Unconstrained Optimisation, in W. Murray (ed.), *Numerical Methods for Unconstrained Optimisation,* Academic Press, London and New York, 1972.

[27] Pratt, M.J., Smoothing of Parametric Surfaces, in *Proc. 3rd Anglo-Hungarian Seminar on Computer Aided Geometric Design,* Cambridge University Engineering Dept., 1983.

[28] Pratt, M.J, Interactive Geometric Modelling for Integrated CAD/CAM, in *Eurographics 84 Tutorial Notes,* Springer Verlag, 1985a.

[29] Pratt, M.J., Parametric Curves and Surfaces as used in Engineering, these Proceedings, 1985b.

[30] Requicha, A.A.G. & Voelcker, H.B., Solid Modelling: a Historical Summary & Contemporary Assessment, IEEE Computer Graphics & Applications, **2**, 2, pp. 9-24, 1982.

[31] Sabin, M.A., Two Basic Interrogations of Parametric Surfaces, Report VTO/MS/148, British Aircraft Corporation, Weybridge, Surrey, 1968.

[32] Sabin, M.A., Techniques of Steplength Control for Marching Method Interrogations, Report VTO/MS/157, British Aircraft Corporation, Weybridge, Surrey, 1969a.

[33] Sabin, M.A., Interrogations involving Two Parametric Surfaces, Report VTO/MS/158, British Aircraft Corporation, Weybridge, Surrey, 1969b.

[34] Sabin, M.A., First Approximations, Starts and Finishes, Report VTO/MS/159, British Aircraft Corporation, Weybridge, Surrey, 1969c.

[35] Sabin, M.A., A Method for displaying the Intersection Curve of two Quadric Surfaces, Comput. J., **19**, pp. 336-338, 1976.

[36] Sabin, M.A., Recursive Subdivision Techniques, these Proceedings, 1985.

[37] Salmon, G., *Modern Higher Algebra*, Hodges, Smith & Co., Dublin, 1866.

[38] Sarraga, R.F., Algebraic Methods for Intersections of Quadric Surfaces in GMSOLID, Computer Vision, Graphics & Image Processing **22**, pp. 222-238, 1983.

[39] Sederberg, T.W., Anderson, D.C. & Goldman, R.N., Implicit Representation of Parametric Curves and Surfaces, Computer Vision, Graphics and Image Processing, **28**, pp. 72-84, 1984.

[40] South, N.E. & Kelly, J.P., Analytic Surface Methods, Internal report, NC Development Unit, Ford Motor Co., Dearborn, Michigan, 1965.

[41] Uspensky, J.V., Theory of Equations, McGraw-Hill, 1948.

[42] Warnock, J.E., A Hidden-line Algorithm for Halftone Picture Representation, University of Utah Computer Science Dept. Report TR 4-5, 1968.

[43] Weiss, R.A., BE VISION, a Package of IBM 7090 FORTRAN Programs to Draw Orthographic Views of Combinations of Plane and Quadric Surfaces, J.ACM, **13**, pp. 194-204, 1966.

[44] Woon, P.Y. & Freeman, H., A Procedure for generating Visible-line Projections of Solids bounded by Quadric Surfaces, in Proc. 1971 IFIP Congress 1120-1125, North-Holland Pub. Co., 1971.

DEFINING AND DESIGNING CURVED FLEXIBLE TENSILE SURFACE STRUCTURES

C.J.K. Williams
(University of Bath)

1. INTRODUCTION

There are many examples of structures which consist of a thin curved surface carrying imposed loads mainly by membrane stresses acting in the plane of the surface. In nature there are egg shells, spider webs and bubbles. Man has made concrete shell roofs, tents, sails and balloons.

Curved surface structures can be classified in many ways, but the most important distinction is between structures which can carry both tensile and compressive membrane stresses and those which can only carry tensile membrane stresses.

The state of stress in any structure modifies its stiffness, tensile stresses increase stiffness and compressive stresses decrease stiffness. In many structures this effect can be neglected but in others such as a violin string in tension or a slender column in compression the effect dominates their behaviour.

The negative stiffness due to compressive stresses is due to the fact that deflections of the structure make it less efficient in carrying the load. A localised inwards load on an otherwise unloaded dome will reduce the local curvature of the structure possibly leading to buckling. When more uniform loads are applied causing compressive membrane stresses, geometric imperfections in the surface become magnified, again possibly leading to buckling. Thus if a curved surface structure is to carry compressive membrane stresses it must have sufficient elastic stiffness to ensure that buckling does not occur. The stiffness is of two types. Firstly there is the in plane resistance to changes of length on the surface and secondly there is resistance to bending of the surface. If the in plane stiffness is relatively high and the bending stiffness relatively low, buckling occurs suddenly after very little deformation and the buckling load is very sensitive to geometric imperfections in the surface. However, if the in plane stiffness is relatively low and the bending stiffness relatively high, much larger deformations can occur before buckling.

The structural action due to in plane membrane stresses is far more efficient than that due to bending and therefore all curved surface structures carrying loads which cause compressive membrane stresses should have reasonably high in plane stiffness. This can either be achieved by forming the surface from a continuous sheet or from discrete members which form a triangular pattern on the surface.

The positive stiffness due to tensile stresses is due to the fact that deflections of the structure make it more efficient in carrying the load. This increase in stiffness is known as initial stress or geometric stiffness to distinguish it from elastic stiffness. Compressive stresses produce negative geometric stiffness.

If a surface is prestressed in tension by, for example, inflating a balloon or pulling on the guys of a tent, the surface can carry a variety of loads by changing shape with relatively little change in membrane stress. To achieve the change in shape the surface must be relatively flexible both in the plane of the surface and in bending. This flexibility is of tremendous practical use in demountable structures such as inflatable boats, sails and tents and also in the erection of permanent large scale tensile roofs. These tension roofs may be prestressed against masts, boundary cables etc or by internal air pressure to produce an air-supported structure which is entered through an air lock or revolving doors. The roof membrane can be laid out on the ground or hung from supports and then jacked or inflated into position. The cost of erection is the biggest problem with large concrete shells and steel frameworks which have to be constructed in situ.

The bending stiffness of most prestressed tension membranes is so low that it can be ignored even in the most accurate analysis. This is not the case with surfaces designed to carry compressive membrane stresses where localised loads and strain incompatibilities at boundaries can produce appreciable bending stresses. Any buckling analysis must take into account bending stiffness or the resulting answer for the buckling load will be zero.

Most high strength materials are relatively stiff and therefore the required in plane flexibility of prestressed membrane structures is achieved by using only two sets of crossing cables in a cable net or only two sets of crossing yarns (the warp and weft yarns) in a woven coated fabric. In plane strains can then occur by relative rotation of the two sets of members. A cable net can therefore be fabricated flat with all the members in tension and erected to form a doubly curved surface. It will be seen later than the relative rotation of the cables depends only upon the Gaussian curvature of the surface. The amount of relative rotation of the yarns of a coated fabric is more limited and therefore the fabric panels are usually shaped before being sewn, welded or glued together. Fabric is normally woven in strips of perhaps 2 m width.

The coating is applied to protect the base fabric and to make the fabric airtight. Many different combinations of base fabric and coating are used. Most large scale structures covering sports stadia, exhibition halls, etc are made of woven glass fibre coated with PTFE. A cable net may be clad with a fabric, opaque, translucent or transparent panels or even with timber boarding and tiles.

As a tension membrane structure deflects under load its stiffness will normally increase as the average tension in the membrane increases. The prestress is chosen to give sufficient stiffness under low loads and avoid loss of tension under most working conditions. Localised loss of tension in one direction in the surface can often be accepted under extreme conditions leading to wrinkling of the surface without damaging it. The lower the prestress, the lower the membrane tensions, foundation loads, etc and therefore the lower the cost.

The main problem in the structural analysis under load is not that of analysing the membrane itself but of analysing the load on it which often changes as the structures deflect . The snow load on a flexible roof is influenced by deflection of the roof. Wind load is never constant and its effect should therefore be studied using a dynamic analysis. The mass of the membrane is usually dominated by the 'added' or 'virtual' mass of the air that must move with the membrane. This produces a pressure difference across the membrane proportional to the acceleration of the membrane and dependent upon the scale of the deformation. Wind also modifies the stiffness of the membrane as can be seen by considering wind flowing over a flat membrane. If the membrane deflects slightly the wind must speed up over peaks and slow down over troughs. This causes the air pressure to decrese over the peaks and increase over the troughs thus tending to increase the deflection which is equivalent to a negative stiffness. This may lead to a loss of static stability, known as divergence. Wind may also produce damping like terms, which if negative can lead to a loss of dynamic stability, known as flutter.

The discontinuity in the tangential component of air speed either side of a membrane is equivalent to a sheet vortex. If the air flow is assumed to be otherwise irrotational, the classical methods of aeroelasticity can be used to predict the pressure difference across the surface in terms of its displacement, velocity etc. However, wind flow will usually separate from the structure so that the assumption is not valid. Aeroelasticity applied to aircraft wings normally relies on the Kutta condition which states that the airflow leaves the trailing edge smoothly. There is no equivalent condition for membrane structures except, perhaps, hand gliders and sails under certain conditions.

The author discussed these problems in two papers at a recent conference [9], [10] devoted exclusively to air supported structures and came to the conclusion that the only solution is to use wind tunnel tests using aeroelastic models which scale the elastic stiffness and geometric stiffness of the full scale structure. At the same conference Barnes [1], [2] described the matrix and relaxation methods that can be used for nonlinear structural analysis assuming the loads to be known.

Fabric structures are prone to failure by tearing and tear resistance is very often more important than tensile strength in the choice of fabric. The same applies to cable nets but to a lesser extent. Williams and Gaafar [7] give results which predict tear resistance in terms of yarn and coating properties.

Having discussed flexible prestressed membrane structures in general, the remainder of the paper will consider their geometry under prestress and how this is controlled by the requirements of static equilibrium. Before we can do this we must first establish the basic geometric and equilibrium relationships.

2. DIFFERENTIAL GEOMETRY

The contents of this section will be familiar to many readers. It is included for completeness and to show the similarity between the geometric relationships and the equilibrium relationships to be discussed in section 3.

The structural grid of most tension membranes is sufficiently fine to be approximated to a continuum. Figure 1 shows a portion of surface defined by the position vector

$$\begin{aligned} \mathbf{r} &= \mathbf{r}(\Theta^1,\Theta^2) \\ &= x^1(\Theta^1,\Theta^2)\mathbf{i}_1 + x^2(\Theta^1,\Theta^2)\mathbf{i}_2 + x^3(\Theta^1,\Theta^2)\mathbf{i}_3 \qquad (2.1) \\ &= x^k\mathbf{i}_k \end{aligned}$$

by the summation convention where k is given the values 1,2 and 3 and the results summed. x^k are cartesian coordinates and $\mathbf{i}_k$ unit vectors in the direction of increasing x^k. Vectors will always be denoted by bold type. Lines of constant Θ^1 or Θ^2 can be drawn on the surface and in general they will not cross at right angles nor will there be a constant spacing between intersections. Θ^1 and Θ^2 form a system of coordinates on the surface.

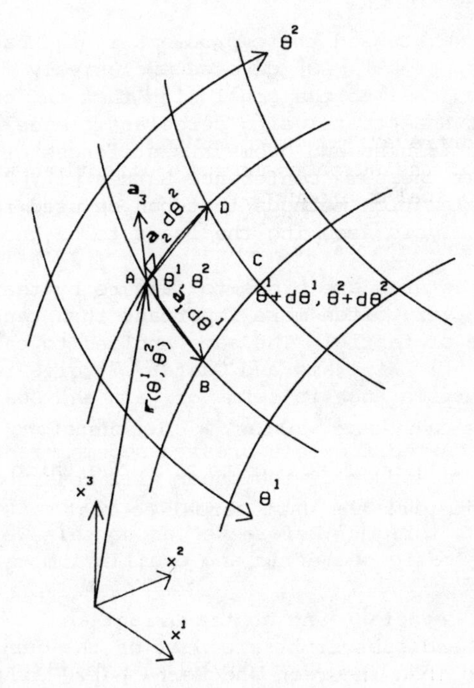

Figure 1

It has already been stated that most flexible tension membrane structures have a two way structural grid composed of two sets of crossing cables or yarns of fabric. It will therefore usually be convenient for the coordinate curves Θ^1 or Θ^2 = constant to follow the structural grid.

At each point on the surface there are two sets of base vectors. Firstly there are the covariant base vectors

$$\mathbf{a}_\alpha = \frac{\partial \mathbf{r}}{\partial \Theta^\alpha} \tag{2.2}$$

which are also written $\mathbf{r}_{,\alpha}$ where, α means partial differentiation with respect to Θ^α.

In this equation α may have the value 1 or 2 and this applies to all Greek indices. Latin indices, such as the k in (2.1) may have the values 1,2 or 3. $\mathbf{a}_1$ and $\mathbf{a}_2$ lie in the local plane of the surface in the directions of increasing Θ^1 and Θ^2 respectively. In general $\mathbf{a}_1$ and $\mathbf{a}_2$ are not unit vectors. The third covariant base vector, $\mathbf{a}_3$, is the local unit normal to the surface

$$\mathbf{a}_3 = \frac{\mathbf{a}_1 \times \mathbf{a}_2}{|\mathbf{a}_1 \times \mathbf{a}_2|}. \tag{2.3}$$

The two contravariant base vectors, $\mathbf{a}^1$ and $\mathbf{a}^2$, lie in the local plane of the surface and have direction and magnitude such that the scalar product

$$\begin{aligned} \mathbf{a}^\alpha . \mathbf{a}_\beta &= \delta^\alpha_\beta \\ &= 1 \text{ if } \alpha = \beta \\ &= 0 \text{ if } \alpha \neq \beta. \end{aligned} \tag{2.4}$$

δ^β_α are the Kronecker deltas. $\mathbf{a}^1$ is therefore perpendicular to $\mathbf{a}_2$ and $\mathbf{a}^2$ is perpendicular to $\mathbf{a}_1$. The third contravariant vector, $\mathbf{a}^3$, is also the unit normal so that

$$\mathbf{a}^3 = \mathbf{a}_3. \tag{2.5}$$

The terms covariant and contravariant and the use of superscripts and subscripts are part of the curvilinear tensor notation described in Green and Zerna [3]. Quantities have superscripts or subscripts and sometimes both according to how they vary under a change of coordinates. Brackets will be used when there is a danger of confusing a superscript and a power. The quantities $a_{\alpha\beta}$ and $a^{\alpha\beta}$ defined by the scalar products

$$\begin{aligned} a_{\alpha\beta} &= a_{\beta\alpha} = \mathbf{a}_\alpha . \mathbf{a}_\beta \\ a^{\alpha\beta} &= a^{\beta\alpha} = \mathbf{a}^\alpha . \mathbf{a}^\beta \end{aligned} \tag{2.6}$$

are the covariant and contravariant metric surface tensors respectively. They are second order tensors in that they each have two indices. It follows from (2.4) and (2.6) that the base vectors can be written in terms of each other

$$\mathbf{a}_\alpha = a_{\alpha\beta}\mathbf{a}^\beta \tag{2.7}$$

$$\mathbf{a}^\alpha = a^{\alpha\beta}\mathbf{a}_\beta .$$

The distance, ds, between two adjacent points Θ^1,Θ^2 and $\Theta^1+d\Theta^1$, $\Theta^2+d\Theta^2$ on the surface is given by

$$ds^2 = d\mathbf{r}.d\mathbf{r} = (\mathbf{a}_\alpha d\Theta^\alpha).(\mathbf{a}_\beta d\Theta^\beta)$$

$$= a_{\alpha\beta}d\Theta^\alpha d\Theta^\beta \tag{2.8}$$

which is known as the first fundamental form of the surface.

The quantity, a, is defined as

$$a = a_{11}a_{22} - (a_{12})^2 \tag{2.9}$$

and $\sqrt{a}\, d\Theta^1 d\Theta^2$ is the area on the surface bounded by the co-ordinate lines through Θ_1,Θ_2 and through $\Theta^1+d\Theta^1,\Theta^2+d\Theta^2$.

$b_{\alpha\beta}$ is defined by the scalar product

$$b_{\alpha\beta} = b_{\beta\alpha} = \mathbf{a}_3 . \frac{\partial^2\mathbf{r}}{\partial\Theta^\alpha\partial\Theta^\beta} = \mathbf{a}_3.\mathbf{r}_{,\alpha\beta} = \mathbf{a}_3.\mathbf{a}_{\alpha,\beta} = -\mathbf{a}_\alpha.\mathbf{a}_{3,\beta} \tag{2.10}$$

and the scalar product

$$d\mathbf{r}.d\mathbf{a}_3 = -\, b_{\alpha\beta}d\Theta^\alpha d\Theta^\beta \tag{2.11}$$

is known as the second fundamental form of the surface.

Associated tensors are formed by multiplying by $a_{\alpha\beta}$ or $a^{\alpha\beta}$ and adding by the summation convention. Thus, for example

$$A^{\mu\alpha}_{\cdot\cdot\rho}\, a_{\alpha\beta} = A^{\mu}_{\cdot\beta\rho}$$

$$A^{\mu}_{\cdot\beta\rho}\, a^{\beta\lambda} = A^{\mu\lambda}_{\cdot\cdot\rho}$$

The dots are used to maintain the order of the indices. The above two equations are consistent since

$$a_{\alpha\beta}a^{\beta\lambda} = \delta^\lambda_\alpha \tag{2.12}$$

which can be solved to give

$$a^{11} = \frac{a_{22}}{a}\,,\; a^{22} = \frac{a_{11}}{a}\,,\; a^{12} = -\frac{a_{12}}{a}\,. \tag{2.13}$$

Now

$$b^{\alpha}_{\cdot\beta} = b_{\lambda\beta}a^{\lambda\alpha}$$

but since both $a^{\lambda\alpha}$ and $b_{\lambda\beta}$ are symmetric ($a^{\alpha\lambda}=a^{\lambda\alpha}$, $b_{\beta\lambda}= b_{\lambda\beta}$)

$$b^{\alpha}_{\cdot\beta} = b_{\beta}^{\cdot\alpha}$$

so that in this case the dot is not required and we can write b^{α}_{β}.

The normal, $\mathbf{a}_3$, is a unit vector and therefore $d\mathbf{a}_3$ lies in the plane of the surface. Thus since $d\mathbf{r} = \mathbf{a}_{\alpha}d\theta^{\alpha}$, (2.11) shows that

$$d\mathbf{a}_3 = - b^{\lambda}_{\beta}\mathbf{a}_{\lambda}d\theta^{\beta}. \tag{2.14}$$

The normal curvature at a point on the surface is equal to

$$-\frac{d\mathbf{r}\,.\,d\mathbf{a}_3}{d\mathbf{r}\,.\,d\mathbf{r}} = \frac{b_{\alpha\beta}d\theta^{\alpha}d\theta^{\beta}}{a_{\lambda\rho}d\theta^{\lambda}d\theta^{\rho}} \tag{2.15}$$

in the direction specified by the ratio $\frac{d\theta^2}{d\theta^1}$. The vectors $d\mathbf{r}$ and $d\mathbf{a}_3$ are parallel if

$$d\mathbf{a}_3 + k\,d\mathbf{r} = 0$$

and thus if

$$\mathbf{a}_{\alpha}\,.\,(b^{\lambda}_{\beta}\mathbf{a}_{\lambda} - k\,\mathbf{a}_{\beta})\;d\theta^{\beta} = 0$$

or

$$(b_{\alpha\beta} - k\,a_{\alpha\beta})d\theta^{\beta} = 0\,. \tag{2.16}$$

(2.16) only has a non-trivial solution, $d\theta^{\beta} \neq 0$ if the determinant $|b_{\alpha\beta} - ka_{\alpha\beta}| = 0$. This results in a quadratic equation for k and the corresponding two values of the ratio

$\frac{d\Theta^2}{d\Theta^1}$ also give the minimum and maximum values of the normal curvature (2.15) which is then equal to the appropriate value of k. The two values of k are the principal curvatures and it is easy to show that they always occur in orthogonal directions (unless the two values of k are equal in which case all directions are principal curvature directions).

The mean curvature, H, is the mean of the two values of k,

$$2H = \frac{a_{11}b_{22} - 2a_{12}b_{12} + a_{22}b_{11}}{a_{11}a_{22} - (a_{12})^2} = b^{\alpha}_{\alpha} \tag{2.17}$$

and the Gaussian curvature, K, is the product of the two values

$$K = \frac{b_{11}b_{22} - (b_{12})^2}{a_{11}a_{22} - (a_{12})^2} = b^1_1 b^2_2 - b^2_1 b^1_2 = \frac{1}{2}\,(b^{\alpha}_{\alpha}b^{\beta}_{\beta} - b^{\beta}_{\alpha}b^{\alpha}_{\beta}). \tag{2.18}$$

The Christoffel symbols of the second kind can be defined as

$$\Gamma^{\lambda}_{\alpha\beta} = \Gamma^{\lambda}_{\beta\alpha} = \mathbf{a}^{\lambda} . \mathbf{a}_{\alpha,\beta} \tag{2.19}$$

(note: Green and Zerna [3] put a $^-$ over the symbol when applied to surface theory).

$\Gamma^{\lambda}_{\alpha\beta}$ can be expressed in terms of the metric surface tensor:

$$\Gamma^{\lambda}_{\alpha\beta} = a^{\lambda\rho}\mathbf{a}_{\rho}.\mathbf{a}_{\alpha,\beta}$$

$$= \frac{1}{2}\, a^{\lambda\rho}\,(a_{\rho\beta,\alpha} + a_{\rho\alpha,\beta} - a_{\alpha\beta,\rho}). \tag{2.20}$$

However the Christoffel symbols are not tensors since they do not obey the appropriate rules under a transformation of coordinates.

(2.10) and (2.19) produce the Gauss equations:

$$\mathbf{a}_{\alpha,\beta} = \Gamma^{\lambda}_{\alpha\beta}\mathbf{a}_{\lambda} + b_{\alpha\beta}\mathbf{a}_3 \tag{2.21}$$

as can be seen by scalar multiplying (2.21) by $\mathbf{a}^{\rho}$ or $\mathbf{a}_3$.

Setting $\beta = 1$ in (2.21) and differentiating with respect to Θ^2 must give the same result as setting $\beta = 2$ and differentiating with respect to Θ^1. Scalar multiplying by $\mathbf{a}_3$ then gives

$$\Gamma^{\lambda}_{\alpha 1}\, b_{\lambda 2} + b_{\alpha 1,2} = \Gamma^{\lambda}_{\alpha 2}\, b_{\lambda 1} + b_{\alpha 2,1}\,. \tag{2.22}$$

Scalar multiplying by $\mathbf{a}_{\mu}$ instead of $\mathbf{a}_{3}$ gives

$$0 = 0 \text{ if } \mu = \alpha$$

and

$$K = \frac{b_{11}b_{22}-(b_{12})^2}{a_{11}a_{22}-(a_{12})^2} = \frac{1}{a_{11}a_{22}-(a_{12})^2}\left[a_{12,12} - \frac{1}{2}(a_{11,22}+a_{22,11}) - \Gamma^{\lambda}_{11}\Gamma^{\rho}_{22}a_{\lambda\rho} + \Gamma^{\lambda}_{12}\Gamma^{\rho}_{12}a_{\lambda\rho}\right] \tag{2.23}$$

if $\mu \neq \alpha$.

The two equations (2.22) ($\alpha = 1$ or 2) are the Codazzi equations and (2.23) is Gauss' Theorem. Gauss' Theorem is particularly important since it effectively shows that the Gaussian curvature can be expressed in terms of the metric tensor and its derivatives only. In particular it shows that for a surface to deform in such a way that the Gaussian curvature does not remain constant, certain (but not necessarily all) lengths on the surface must change.

Lastly the fundamental theorem of surface theory states that $a_{\alpha\beta}$ and $b_{\alpha\beta}$ uniquely determine a surface except for its position and orientation in space. The fundamental theorem can be proved using physical arguments by choosing the position and orientation of $\mathbf{a}_1$ and $\mathbf{a}_2$ at a point on the surface and 'growing' the remainder of the surface away from that point. $a_{\alpha\beta}$ and $b_{\alpha\beta}$ must, of course, satisfy the two Codazzi equations and Gauss' theorem which are effectively compatibility equations which ensure that the surface 'fits together'.

3. EQUILIBRIUM RELATIONSHIPS

Figure 2 shows a short imaginary cut in a surface. The orientation of the cut is specified by the vector, $d\mathbf{v}$, which lies in the local plane of the surface perpendicular to the cut. The length of the cut is equal to the magnitude of $d\mathbf{v}$.

If the surface contains membrane stresses these will be a force, $d\mathbf{f}$, crossing the cut. $d\mathbf{f}$ will lie in the local plane of the surface since we are assuming that bending stiffness is zero so that there are no shear forces perpendicular to the surface.

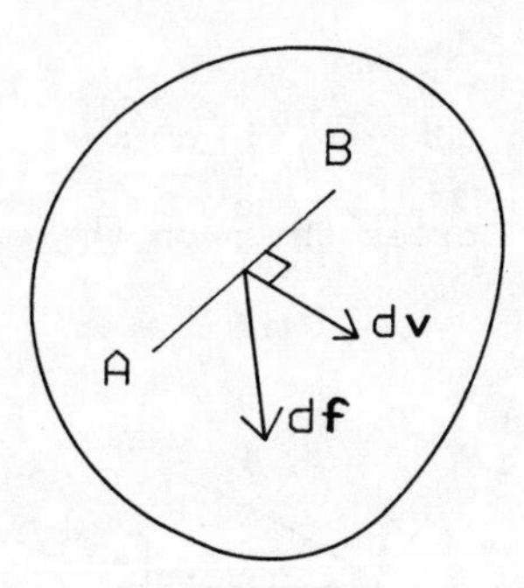

Figure 2

$d\mathbf{f}$ clearly must depend on the direction and magnitude of $d\mathbf{v}$ and therefore if

$$d\mathbf{v} = dv^{\alpha}\mathbf{a}_{\alpha} = dv_{\alpha}\mathbf{a}^{\alpha} \tag{3.1}$$

we can write

$$d\mathbf{f} = df^{\alpha}\mathbf{a}_{\alpha} = df_{\alpha}\mathbf{a}^{\alpha}$$

$$= n^{\alpha\beta}dv_{\alpha}\mathbf{a}_{\beta}. \tag{3.2}$$

$n^{\alpha\beta}$ are the membrane stress resultants and are a second order tensor.

If A on Figure 2 is at Θ^1, Θ^2 and B at $\Theta^1 + d\Theta^1, \Theta^2 + d\Theta^2$, the vector AB is equal to $\mathbf{a}_{\lambda}d\Theta^{\lambda}$. Then

$$dv_{\alpha}\mathbf{a}^{\alpha} = \in_{\alpha\beta}\mathbf{a}^{\alpha}d\Theta^{\beta} \tag{3.3}$$

where the $\in$ tensors are equal to

$$\in_{12} = -\in_{21} = \sqrt{a} \tag{3.4}$$

$$\in_{11} = \in_{22} = 0$$

and a is given by (2.9). (3.3) can be proved by performing the scalar product of (3.3) and $\mathbf{a}_{\lambda}d\Theta^{\lambda}$ and obtaining zero and performing the vector product between the same two vectors and obtaining $a_{\alpha\beta}d\Theta^{\alpha}d\Theta^{\beta}\,\mathbf{a}_3$.

(3.2) and (3.3) produce

$$df = n^{\alpha\beta} \in_{\alpha\lambda} d\Theta^{\lambda} \mathbf{a}_{\beta} \quad (3.5)$$

which results in the forces shown on the small quadrilateral in Figure 3.

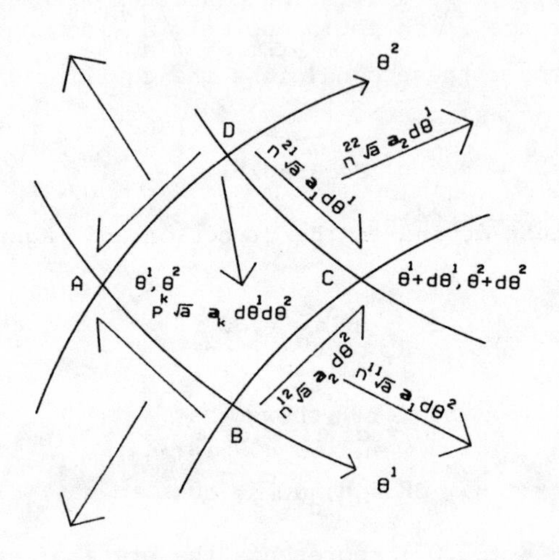

Figure 3

$p^k \mathbf{a}_k$ is the external load including own weight, inertia forces etc. applied to the membrane per unit surface area. The membrane force per unit length in the surface can be calculated from $n^{\alpha\beta}$ but is not numerically equal to $n^{\alpha\beta}$ since in general a, a_{11} and a_{22} are not equal to 1.

Taking moments about the normal to the surface shows that $n^{\alpha\beta}$ is symmetric ($n^{\alpha\beta} = n^{\beta\alpha}$) and adding all the forces on the element in figure 3 results in

$$\frac{\partial}{\partial\Theta^{\alpha}} \left[n^{\alpha\beta} \sqrt{a}\, \mathbf{a}_{\beta} \right] + p^k \sqrt{a}\, \mathbf{a}_k = 0 . \quad (3.6)$$

This vector equilibrium equation is equivalent to 3 scalar equations obtained by scalar multiplying by $\mathbf{a}^{\lambda}$ or $\mathbf{a}_3$:

$$n^{\alpha\lambda}{}_{,\alpha} + n^{\alpha\lambda} \Gamma^{\beta}_{\alpha\beta} + n^{\alpha\beta} \Gamma^{\lambda}_{\alpha\beta} + p^{\lambda} = 0 \quad (3.7)$$

$$n^{\alpha\beta} b_{\alpha\beta} + p^3 = 0 \; . \qquad (3.8)$$

4. UNLOADED RECIPROCAL SURFACES

In many applications we will be concerned with the state of stress in a surface which is loaded only at its boundaries so that $p^k = 0$. Under these conditions the equilibrium equation (3.6) is satisfied by

$$n^{1\beta}\sqrt{a}\,\mathbf{a}_\beta = \mathbf{R},_2$$

$$n^{2\beta}\sqrt{a}\,\mathbf{a}_\beta = -\mathbf{R},_1$$

or

$$\mathbf{R},_\alpha = \in_{\lambda\alpha} n^{\lambda\beta}\,\mathbf{r},_\beta \, . \qquad (4.1)$$

Comparison of (3.5) and (4.1) shows that

$$d\mathbf{R} = \mathbf{R},_\alpha d\Theta^\alpha = d\mathbf{f} \; . \qquad (4.2)$$

The new surface $\mathbf{R}\,(\Theta^1,\Theta^2)$ represents the state of stress in the surface $\mathbf{r}\,(\Theta^1,\Theta^2)$ in that the force across the cut along $\mathbf{a}_\alpha d\Theta^\alpha$ on $\mathbf{r}$ is represented in both magnitude and direction by the vector $\mathbf{A}_\alpha d\Theta^\alpha$ $(= \mathbf{R},_\alpha d\Theta^\alpha)$ on $\mathbf{R}$. Quantities on $\mathbf{R}$ will be represented by capital letters corresponding to the small letters on $\mathbf{r}$.

The two surfaces have the same unit normal and therefore

$$d\mathbf{A}_3 = d\mathbf{a}_3$$

so that using (2.14)

$$B_{\eta\mu}\mathbf{A}^\eta = b_{\eta\mu}\mathbf{a}^\eta . \qquad (4.3)$$

Scalar multiplying (4.1) and (4.3)

$$B_{\alpha\mu} = \in_{\lambda\alpha} n^{\lambda\beta} b_{\beta\mu} \, . \qquad (4.4)$$

Writing out the four equations (4.4) in full, eliminating n^{11} and n^{22} and inserting the condition $n^{12} = n^{21}$ results in

$$b_{11}B_{22} + b_{22}B_{11} = 2\, b_{12}B_{12}. \tag{4.5}$$

The symmetry of this relationship shows that if **R** represents a state of stress in the unloaded prestressed surface **r**, then **r** also represents a possible state of stress in the unloaded prestressed surface **R**. There is therefore a reciprocal relationship. It should be noted that an unloaded surface can have an infinite number of different states of stress so that for any **r** there is an infinite choice of **R** and vice versa. The two reciprocal surfaces are not, however, arbitrary since equation (4.5) must be satisfied.

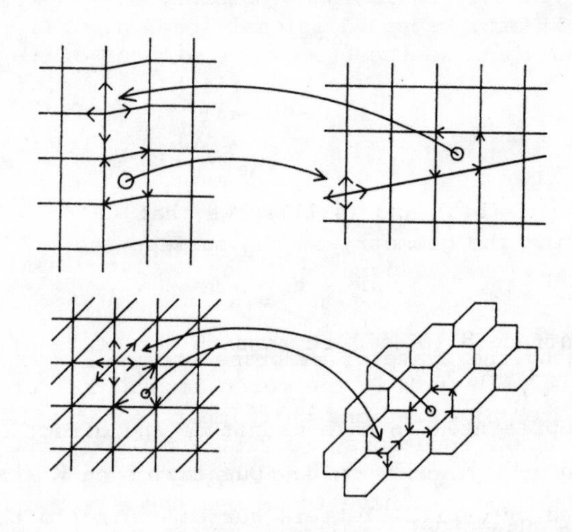

Figure 4

Reciprocal surfaces will be used in some of the work that follows since in studying the surface, **r**, it may be easier to examine **R** instead. Reciprocal surfaces do not have to be continua or even approximate to continua. Figure 4 shows part of two pairs of coarse prestressed reciprocal nets in static equilibrium. The nets do not need to be flat or have only two sets of cables. The equilibrium of a node on one net is ensured by a closed polygon of forces on the other net. Every node on one net corresponds to a polygon on the other. The polygons are traversed clockwise on one net and anticlockwise on the other. Readers may notice the similarity with Bow's notation used for the analysis of plane trusses.

5. THE FORM FINDING OF TENSILE SURFACE STRUCTURES

In the classical membrane theory of shells all geometric quantities and applied loads are assumed to be known. The problem is therefore to solve the three equilibrium equations represented by (3.6) for the three unknowns $n^{\alpha\beta}(=n^{\beta\alpha})$. The problem is thus statically determinate provided that the boundary conditions are of a suitable form.

In the form finding of tensile surface structures the applied loads are known since they consist of only dead weight and inflation pressure (where appropriate). This is because the form finding considers a (possibly imaginary) state before wind, snow, etc. cause additional loading. The geometry will not be known and we therefore have a total of nine unknowns,

$$a_{\alpha\beta} = a_{\beta\alpha}$$

$$b_{\alpha\beta} = b_{\beta\alpha}$$

representing the geometry of the surface and

$$n^{\alpha\beta} = n^{\beta\alpha}$$

representing the state of membrane stress.

There are six equations which must be satisfied. They are the two Codazzi equations, (2.22), and Gauss' theorem, (2.23), which are purely geometric and the three equilibrium equations, (3.6). We therefore have to impose three more conditions and appropriate boundary conditions before the surface and the state of prestress are fully defined.

If a two way cable net is sufficiently fine to be approximated to a continuum and the surface coordinates, Θ^{α}, are chosen to follow the cable directions, then examination of Figure 3 shows that $n^{12}(=n^{21})=0$. The same condition can be applied to a woven fabric structure where it is advisable that the membrane stress under prestress conditions follows the directions of the yarns. This is to avoid large and unpredictable strains in the bias direction of the fabric. Thus in the remainder of this paper we will apply the condition $n^{12} = 0$ which means that only two further conditions need to be applied.

In the case of an unloaded net which is prestressed only by forces applied at its boundary, substitution of $p^k = 0$ and $n^{12} = 0$ into the equilibrium equation (3.6) produces

$$n^{11}\sqrt{a}\,b_{11} = -\,n^{22}\sqrt{a}\,b_{22} = Q\,b_{11}b_{22} \tag{5.1}$$

where

$$\frac{-Q,_1}{Q} = \frac{b_{22,1}}{b_{22}} + \frac{(\frac{1}{2}a_{11,1}a_{22} - a_{12,1}a_{12} + \frac{1}{2}a_{11,2}a_{12})}{a_{11}a_{22} - (a_{12})^2}$$

$$+\frac{b_{11}}{b_{22}}\,\frac{(-a_{12,2}a_{22} + \frac{1}{2}a_{22,1}a_{22} + \frac{1}{2}a_{22,2}a_{12})}{a_{11}a_{22} - (a_{12})^2}$$

(5.2)

$$\frac{-Q,_2}{Q} = \frac{b_{11,2}}{b_{11}} + \frac{(\frac{1}{2}a_{22,2}a_{11} - a_{12,2}a_{12} + \frac{1}{2}a_{22,1}a_{12})}{a_{11}a_{22} -)a_{12})^2}$$

$$+\frac{b_{22}}{b_{11}}\,\frac{(-a_{12,1}a_{11} + \frac{1}{2}a_{11,2}a_{11} + \frac{1}{2}a_{11,1}a_{12})}{a_{11}a_{22} - (a_{12})^2}.$$

Q can be eliminated from the two equations (5.2) by differentiating and subtracting. This leads to one equation in geometric quantities only.

If the cables (and therefore coordinate curves) follow asymptotic directions on the surface, that is directions in which the normal curvature is zero, then (2.15) shows that $b_{11} = b_{22} = 0$. Substituting into the equilibrium equation shows that equilibrium in the normal direction is automatically satisfied provided that the net is unloaded in the normal direction. This leaves two equilibrium equations which can always be satisfied by n_{11} and n_{22}.

6. EQUAL MESH CABLE NETS

Equal mesh cable nets are fabricated flat under a constant tension with a constant internode spacing. A spacing of 0.5m is often used as this enables the net to be walked on after erection for the fixing of cladding etc. without too much danger of falling through. The best known equal mesh net structure is the roof over the stands of the Munich Olympic Stadium. Sometimes the grid is much smaller such as that of the Munich Zoo Aviary which is constructed of stainless steel wires which are crimped by passing between two gear wheels. The wires are then woven and kept in place by the crimp.

The usual method of form finding is to use physical models and non-linear structural analysis computer programs using relaxation techniques or matrix methods. Barnes [1],[2] describes the various methods used. In the form finding process adjustments are made to the boundary geometry and tensions until an acceptable form is obtained. After each adjustment the structural analysis has to be performed to bring the net into equilibrium and the analysis must also ensure equilibrium of masts and curved boundary cables. The main problem is one of data organisation in that nodes may have to be added or removed during the form finding. Nodes are normally numbered in structural analysis programs, but for cable nets it is often easier to number each continuous length of cable so that nodes have two numbers which are effectively the values of Θ^1 and Θ^2.

This method of form finding requires no knowledge of differential geometry since the cartesian coordinates of each node are used as the variables, automatically ensuring that the net fits together. Cable lengths are calculated from the coordinates, then cable tensions using the cables' elastic stiffness. The resultant force on each node is calculated and the nodes moved to try and reduce the resultant forces. This process is repeated until an acceptable state of equilibrium is reached.

The application of differential geometry to equal mesh nets is facilitated by neglecting the different elastic extension of the various members due to different tensions. This is acceptable for preliminary studies, but the changes in length cannot be neglected when producing dimensions for fabrication. In general equal mesh nets are sufficiently fine for the continuum approximation required by differential geometry to be made. Neglecting elastic extension

$$a_{11} = a_{22} = 1 \qquad (6.1)$$

which are the remaining two conditions required so that an equal mesh net can be determined by suitable boundary conditions conditions.

Using (2.20) the Christoffel symbols of the second kind can be written down:

$$\Gamma^1_{11} = \frac{-a_{12}a_{12,1}}{a} \qquad \Gamma^2_{22} = -\frac{a_{12}a_{12,2}}{a}$$

$$\Gamma^1_{22} = \frac{a_{12,2}}{a} \qquad \Gamma^2_{11} = \frac{a_{12,1}}{a} \tag{6.2}$$

$$\Gamma^1_{12} = \Gamma^2_{12} = 0 .$$

Substituting into the Codazzi equations, (2.22), gives

$$-a_{12}a_{12,1}b_{12} + a_{12,1}b_{22} + (1-(a_{12})^2)(b_{11,2} - b_{12,1}) = 0$$

$$-a_{12,}a_{12,2}b_{12} + a_{12,2}b_{11} + (1-(a_{12})^2)(b_{22,1} - b_{12,2}) = 0 \tag{6.3}$$

and into Gauss' theorem, (2.23):

$$K = \frac{b_{11}b_{22} - (b_{12})^2}{1-(a_{12})^2} = -\frac{1}{\sin\alpha}\frac{\partial^2\alpha}{\partial\theta^1\partial\theta^2}, \tag{6.4}$$

where α is the angle between the direction of increasing θ^1 and increasing θ^2. Struik [4] attributes this last result to Tschebycheff [5].

In many cases the own weight of a cable net is so low that it can be neglected during form finding and added later in the analysis as an applied load.

If this is the case and the structure is not inflated, equation (5.2) can be used, which upon substituting $a_{11} = a_{22} = 1$ produces

$$\left[\frac{\dfrac{b_{22,1}}{b_{22}} - \dfrac{b_{11}}{b_{22}}a_{12,2}}{1-(a_{12})^2}\right]_{,2} = \left[\frac{\dfrac{b_{11,2}}{b_{11}} - \dfrac{b_{22}}{b_{11}}a_{12,1}}{1-(a_{12})^2}\right]_{,1} . \tag{6.5}$$

The two Codazzi equations, (6.3), Gauss' theorem, (6.4) and the equilibrium equation, (6.5), are four equations in the four unknowns a_{12}, b_{11}, b_{22} and b_{12}. Unfortunately, the author has not been able to find a general solution to these equations. These are, however, a number of special cases.

Firstly, it has already been noted that any unloaded prestressed net with the cables in the asymptotic directions on the surface must automatically be in equilibrium. Setting $b_{11} = b_{22} = 0$ in the Codazzi equations, (6.3), for an equal mesh net leads to

$$K = \frac{-(b_{12})^2}{1-(a_{12})^2} = -c^2 = \text{constant} . \tag{6.6}$$

There are many surfaces with constant negative Gaussian curvature. Amongst these in the surface of revolution known as the pseudosphere with pseudoradius c (see Struik [4]).

A second special case can be obtained by observing that an unloaded equal mesh net must have a reciprocal net with constant tensions. A special case of a constant tension net is a net where all cables are straight. If such a net exists, it must be uniquely defined by three cables from one set provided that no two of the three cables lie in the same plane. The three cables define the net because when we come to add the second set of cables we find that there can only be one straight line drawn through a point on one of the three original cables which also touches the other two.

The theory of homogeneous linear equations shows that we can always ensure that a surface of the form

$$c_1+c_2x^1+c_3x^2+c_4x^3+c_5x^1x^2+c_6x^2x^3+c_7x^3x^1+c_8(x^1)^2+c_9(x^2)^2+c_{10}(x^3)^2=0 \tag{6.7}$$

passes through any nine points in space by a suitable choice of the ten constants, c_1 to c_{10}, or to be more exact, the nine ratios between the constants. (6.7) represents an ellipsoid or a hyperboloid of one or two sheets. Special cases of these surfaces include the sphere and the hyperbolic paraboloid. If the nine points are chosen such that there are three on each of our original straight cables, the resulting surface will be a hyperboloid of one sheet or a hyperbolic paraboloid. These two surfaces contain two sets of straight line generators and since the only conic section containing three collinear points is a pair of straight lines, each set of three points will lie on a generator.

Thus the only surface containing two sets of straight line generators is the hyperboloid of one sheet which includes the special case of the hyperbolic paranoloid. By a suitable choice of the cartesian axes x^k, a hyperboloid of one sheet can be expressed in the form

$$\mathbf{R} = X^k \mathbf{i}_k$$

$$X^1 = E \left[\frac{\cos(u+v) - 1}{\cos(u-v)} \right] \qquad (6.8)$$

$$X^2 = F \frac{\sin(u+v)}{\cos(u-v)}$$

$$X^3 = G \tan(u-v)$$

where

$$u = u(\Theta^1)$$
$$v = v(\Theta^2).$$

(6.8) includes the hyperbolic paraboloid

$$X^1 = -2Euv$$
$$X^2 = F(u+v) \qquad (6.9)$$
$$X^3 = G(u-v).$$

After differentiating (6.8)

$$\mathbf{R},_1 = \frac{u,_1}{\cos^2(u-v)} [- E \sin 2v\, \mathbf{i}_1 + F \cos 2v\, \mathbf{i}_2 + G\, \mathbf{i}_3] \qquad (6.10)$$

$$\mathbf{R},_2 = \frac{v,_2}{\cos^2(u-v)} [- E \sin 2u\, \mathbf{i}_1 + F\cos 2u\, \mathbf{i}_2 - G\, \mathbf{i}_3]$$

from which it can be seen that the coordinate curves are indeed straight lines.

Thus for the reciprocal equal mesh net, we have from (4.1)

$$\mathbf{r},_1 = \frac{-E \sin 2u\, \mathbf{i}_1 + F \cos 2u\, \mathbf{i}_2 - G\, \mathbf{i}_3}{\sqrt{(E^2 \sin^2 2u + F^2 \cos^2 2u + G^2)}}$$

$$\mathbf{r},_2 = \frac{E \sin 2v\, \mathbf{i}_1 - F \cos 2v\, \mathbf{i}_2 - G\, \mathbf{i}_3}{\sqrt{(E^2 \sin^2 2v + F^2 \cos^2 2v + G^2)}} \qquad (6.11)$$

or

$$\mathbf{r} = \mathbf{p}(\Theta^1) + \mathbf{q}(\Theta^2). \tag{6.12}$$

(6.12) shows that the resulting equal mesh net must be a translation surface formed by sliding the space curve $\mathbf{p}(\Theta^1)$ along $\mathbf{q}(\Theta^2)$ or vice versa.

Setting F = E, u = Θ^1 and v = Θ^2 results in a right helicoid formed by the intersection of two sets of helices:

$$x^1 = \frac{E}{2\sqrt{(E^2+G^2)}} (\cos 2\Theta^1 - \cos 2\Theta^2)$$

$$x^2 = \frac{E}{2\sqrt{(E^2+G^2)}} (\sin 2\Theta^1 - \sin 2\Theta^2) \tag{6.13}$$

$$x^3 = \frac{-G}{\sqrt{(E^2+G^2)}} (\Theta^1 + \Theta^2)$$

or

$$\frac{x^1}{x^2} = \tan\left[\frac{\sqrt{(E^2+G^2)}}{G} x^3\right].$$

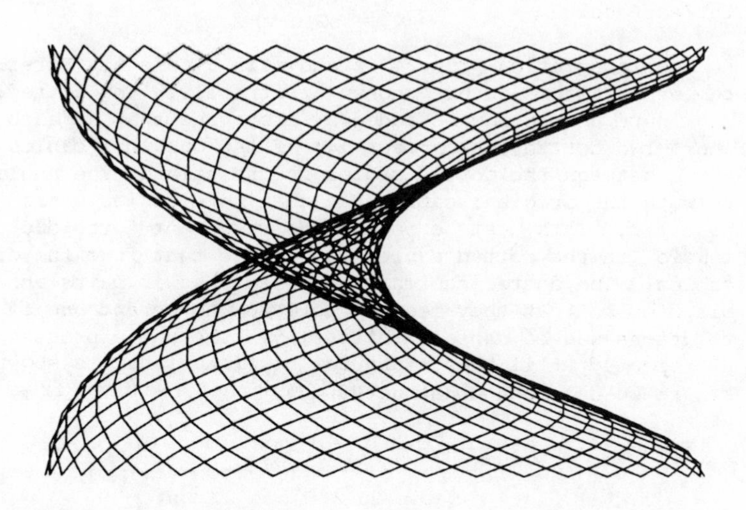

Figure 5. Right Helicoid. Equal mesh net viewed from an angle of 80 degrees to axis.

Figure 5 shows a plot of (6.13). The right helicoid is a minimal surface which means that its mean curvature is zero. However the state of stress in the equal mesh net does not correspond to the uniform surface tension of a soap film.

For the special case of the hyperbolic paraboloid constant tension net, (6.11) becomes

$$\mathbf{r}_{,1} = \frac{-2Eu\,\mathbf{i}_1 + F\,\mathbf{i}_2 - G\,\mathbf{i}_3}{\sqrt{(4E^2u^2+F^2+G^2)}}$$

$$\mathbf{r}_{,2} = \frac{Ev\,\mathbf{i}_1 - F\,\mathbf{i}_2 - G\,\mathbf{i}_3}{\sqrt{(4E^2v^2+F^2+G^2)}} \tag{6.14}$$

so that the two curves forming the translation surface can be any pair of plane curves.

Other prestressed equal mesh nets can be found by integrating equations (6.3), (6.4) and (6.5) numerically. The easiest way to do this is to again consider the reciprocal net which must have constant tensions in all cables to correspond to the equal lengths in the equal mesh net. The cables on the constant tension net must form geodesics on the surface, that is they must appear straight when viewed along the normal to the surface.

The integration process is shown in Figure 6. Firstly the cables in Figure 6a are chosen arbitrarily. The dotted cables on Figure 6b can now be added by observing that at each node where two dotted cables meet two of the original cables, the angle between the dotted cables is the same as the angle between the original cables and that both angles share the same bisector. This must be so for the resultant force due to tension in the dotted cables to balance that from the original cables. The dotted cables must be rotated in pairs about each bisector so that they meet to form new nodes and one of these rotations can be chosen arbitrarily. The whole process can be repeated until the triangular portion of net is shown in Figure 6c has been constructed.

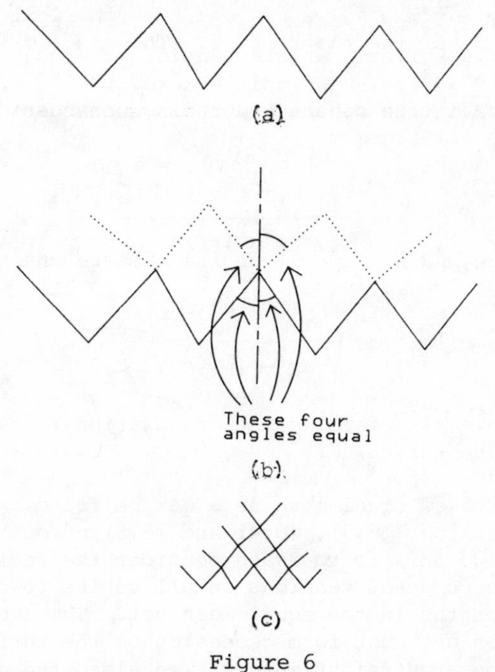

Figure 6

7. PRINCIPAL CURVATURE CABLE NETS

Equal mesh nets have many practical advantages but they have the disadvantage that they cannot be supported at a point such as a mast without producing very high local tensions in the net. The cables should leave a mast radially and some structures consist of radial nets supporting an equal mesh net.

Nets with the cables following directions of principal curvature on the surface will leave a mast radially and they have the advantage that the quadrilaterals formed on the surface are flat which may facilitate the design of cladding. However, as with all radial nets, tensions increase away from masts as does the cable spacing.

The form of principal curvature nets cannot be found by the methods of structural analysis that can be used to produce equal mesh nets. This is because no relationship between cable lengths and tensions will automatically ensure a principal curvature net.

Provided that the net is reasonably fine, we can make the continuum approximation and use differential geometry. If the net is unloaded except by the prestressing force applied at its boundary and by any masts, we can use the spherical image of the surface to determine the surface. The spherical image of any surface, $\mathbf{r}(\Theta^1,\Theta^2)$, is the unit sphere, $\mathbf{s}(\Theta^1,\Theta^2)$, where the coordinate lines on the sphere are drawn such that

$$\mathbf{s}(\Theta^1,\Theta^2) = \frac{\mathbf{r},_1 \times \mathbf{r},_2}{|\mathbf{r},_1 \times \mathbf{r},_2|} = \mathbf{a}_3. \quad (7.1)$$

Thus the surface and its spherical image share the same normal.

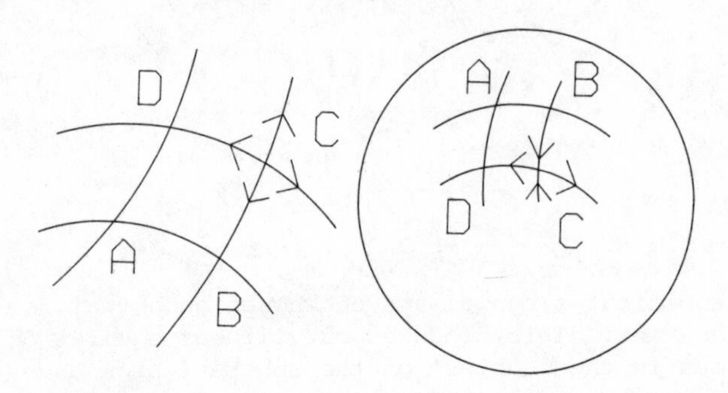

Figure 7

The coordinate lines on the surface and on its spherical image are parallel if and only if the coordinate lines follow the directions of principal curvature on the surface. Examination of Figure 7 shows that if the tensions from the four cables coming into a node are in equilibrium, the forces in the equivalent 'cables' on the spherical image must also be in equilibrium. However, the original surface must have negative Gaussian curvature for its cables to be all in tension. This means that one set of 'cables' on the sphere must be in the opposite order to those on the surface and therefore must be in compression.

Thus we have the result that all unloaded principal curvature nets must correspond to an unloaded orthogonal 'net' on a sphere (the directions of principal curvature are always orthogonal). It should be remembered that both nets have forces applied at their boundaries and may have point loads such as from masts.

On the sphere quantities will be given a $^\circ$ over, thus

$$\mathring{a}_{12} = \mathring{b}_{12} = 0$$

$$\frac{\mathring{b}_{11}}{\mathring{a}_{11}} = \frac{\mathring{b}_{22}}{\mathring{a}_{22}} = -1 \text{ (if the unit normal points outwards).} \tag{7.2}$$

Substituting into the equilibrium equation (5.2) for an unloaded net produces

$$\mathring{Q}\,(\mathring{a}_{22})^{3/2}\,(\mathring{a}_{11})^{\frac{1}{2}} = g(\theta^2)$$

$$\mathring{Q}\,(\mathring{a}_{11})^{3/2}\,(\mathring{a}_{22})^{\frac{1}{2}} = f(\theta^1) \tag{7.3}$$

$$\frac{\mathring{a}_{11}}{\mathring{a}_{22}} = \frac{f(\theta^1)}{g(\theta^2)}.$$

The coordinates can always be chosen such that $f(\theta^1) = g(\theta^2) = T =$ constant. This produces curvilinear squares on the sphere and forces in the 'cables' on the sphere (which have the same magnitude as those on the corresponding principal curvature net)

$$\mathring{n}^{11}\,\ell^3 = -\,\mathring{n}^{22}\,\ell^3 = \frac{T}{\ell} \tag{7.4}$$

where $\ell = \sqrt{\mathring{a}_{11}} = \sqrt{\mathring{a}_{22}}$.

The theory of complex numbers shows that all such patterns of curvilinear squares on the sphere can be obtained by writing

$$\theta^1 + i\theta^2 = \mathcal{Z}(\alpha^1 + i\alpha^2) \tag{7.5}$$

where $\mathcal{Z}$ is any analytic function, $i = \sqrt{-1}$ and the coordinates α^1, α^2 themselves form curvilinear squares on the sphere. The simplest coordinate system α^1, α^2 is

$$x^1 = \frac{\cos\alpha^1}{\cosh\alpha^2}$$

$$x^2 = \frac{\sin\alpha^1}{\cosh\alpha^2} \tag{7.6}$$

$$x^3 = \frac{\sinh\alpha^2}{\cosh\alpha^2}$$

in which α^2 is in the range $-\infty$ to $+\infty$ and α^1 is in range 0 to 2π to cover the sphere once, although there may be times when one wants to cover the sphere more than once which is perfectly acceptable. The net on the sphere corresponding to α^1, α^2 is in equilibrium with two point loads applied along the diameter of the sphere at $\alpha^2 = \pm\infty$.

If $R_{(1)} = \dfrac{a_{11}}{b_{11}}$ and $R_{(2)} = \dfrac{a_{22}}{b_{22}}$ are the principal radii of curvature of the surface that we are trying to obtain, then consideration of the rotation of the normal caused by a displacement on the surface shows that

$$\mathbf{r}_{,1} = R_{(1)}\mathbf{s}_{,1} \tag{7.7}$$

$$\mathbf{r}_{,2} = R_{(2)}\mathbf{s}_{,2}.$$

The requirement that $\mathbf{r}_{,12} = \mathbf{r}_{,21}$ yields

$$\begin{aligned}(R_{(1)}\ell)_{,2} &= R_{(2)}\ell_{,2}\\ (R_{(2)}\ell)_{,1} &= R_{(1)}\ell_{,1}\end{aligned} \tag{7.8}$$

after scalar multiplying by $\mathbf{s}_{,1}$ and $\mathbf{s}_{,2}$ respectively and where again $\ell^2 = \mathring{a}_{11} = \mathring{a}_{22} = \mathbf{s}_{,1}.\mathbf{s}_{,1} = \mathbf{s}_{,2}.\mathbf{s}_{,2}$.

ℓ can be eliminated from 7.8 to give

$$\left[\frac{(R_{(1)} + R_{(2)})_{,2}}{R_{(1)} - R_{(2)}}\right]_{,1} = \left[\frac{(R_{(2)} + R_{(1)})_{,1}}{R_{(2)} - R_{(1)}}\right]_{,2}. \tag{7.9}$$

There are many possible solutions to (7.9) including $R_{(1),1} = R_{(2),1} = 0$ but we will only discuss

$$R_{(1)} + R_{(2)} = \text{constant} = 2c. \tag{7.10}$$

Substituting into (7.7) and (7.8) gives

$$R_{(1)} = c + \frac{e}{\ell^2}$$

$$R_{(2)} = c - \frac{e}{\ell^2}$$

$$a_{11} = \left[c\ell + \frac{e}{\ell}\right]^2 \tag{7.11}$$

$$a_{22} = \left[c\ell - \frac{e}{\ell}\right]^2$$

where e is a constant.

If we now consider a surface $\bar{\mathbf{r}}(\theta^1,\theta^2)$ derived by moving a constant distance h along the normal to the surface represented by (7.10) and (7.11) we obtain

$$\begin{aligned}
\bar{\mathbf{r}} &= \mathbf{r} + h\,\mathbf{s} \\
\bar{\mathbf{r}}_{,1} &= \mathbf{r}_{,1} + h\,\mathbf{s}_{,1} = (R_{(1)} + h)\mathbf{s}_{,1} \\
\bar{\mathbf{r}}_{,2} &= \mathbf{r}_{,2} + h\,\mathbf{s}_{,2} = (R_{(2)} + h)\mathbf{s}_{,2} \\
\bar{R}_{(1)} &= R_{(1)} + h \\
\bar{R}_{(2)} &= R_{(2)} + h \\
\bar{R}_{(1)} + \bar{R}_{(2)} &= 2c + 2h.
\end{aligned} \tag{7.12}$$

Setting h= −c we obtain a minimal surface

$$\bar{R}_{(1)} = -\bar{R}_{(2)} = \frac{e}{\ell^2}$$

$$\bar{a}_{11} = \bar{a}_{22} = \frac{e^2}{\ell^2}\,. \tag{7.13}$$

and (7.4) shows that the state of stress corresponds to a uniform surface tension, T/e. Thus all minimal surfaces are produced by the analytic function, Ƶ, in (7.5) and the constant e. It is well known that minimal surfaces are produced by analytic functions as, for example, in the formulae of Weierstrass (see Struik [4]).

The surface $R_{(1)} + R_{(2)} = 2c$ is a surface parallel to a minimal surface and it has the advantage that it can produce peaks over masts which is not possible with a minimal surface - see Figure 8.

The surface parallel to a minimal surface can be obtained in a number of ways. One method is to investigate the effect of different functions Ƶ in (7.5) and a second is to solve the equation $R_{(1)} + R_{(2)} = 2c$ in the terms of the coordinates

ϕ^1, ϕ^2 such that

$$\begin{aligned} x^1 &= \phi^1 \\ x^2 &= \phi^2 \\ x^3 &= F(\phi^1, \phi^2) . \end{aligned} \tag{7.14}$$

The mean curvature, H, and Gaussian curvature, K, can be obtained as functions of F and its derivatives using (2.17) and (2.18). Hence setting $R_{(1)} + R_{(2)} = \frac{2H}{K} = 2c$ gives a partial differential equation for F. This can be readily solved (given suitable boundary conditions) by the finite difference technique employing relaxation methods. Once the surface has been obtained the directions of principal curvature on the surface can be found and hence the principal curvature net which will automatically be in equilibrium under prestress.

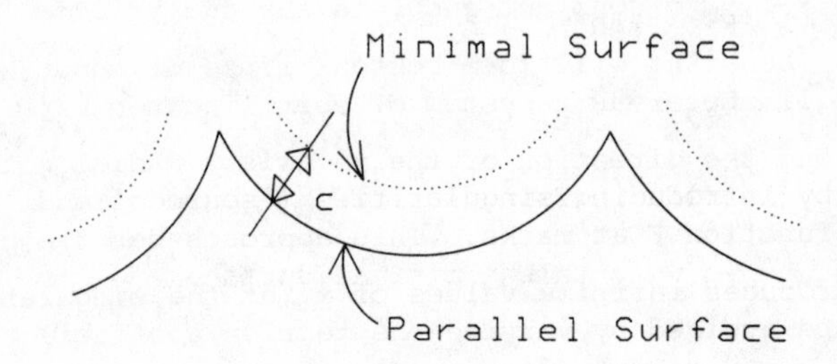

Figure 8

8. CONSTANT TENSION COEFFICIENT NETS

In this section we will again consider nets which are unloaded except by prestress applied at their boundaries. The tension coefficient in a member is defined as the tension in the member divided by its length. Consideration of static equilibrium shows that any structure loaded only at its boundaries and consisting of cables of constant tension coefficient must be in equilibrium if every node (except boundary nodes) has the average of the cartesian coordinates of the nodes to which it is attached by cables. This applies regardless of the form of the structure, it may be a course or fine surface net or a three dimensional structure.

It is also possible to vary the tension coefficients from member to member in which case the cartesian coordinates of a node are the weighted average of those of the nodes to which it is attached. The weighting is the value of tension coefficient in the cable joining the two nodes.

It is very easy to determine the coordinates of such a structure using matrix methods since the coordinates of the nodes are related by linear equations. Alternatively in the case of a fine grid two-way surface net of constant tension coefficient, the requirement of equilibrium becomes simply

$$x^k_{,11} + x^k_{,22} = 0 \tag{8.1}$$

which is Laplace's Equation. The net is therefore defined by any three solutions to Laplace's equation.

One possibility is to write

$$x^1 + ix^2 = F(\Theta^1 + i\Theta^2)$$

$$x^3 = c\,\Theta^1 \tag{8.2}$$

where $i = \sqrt{-1}$, c = constant and F is any analytic function. Lines Θ^1 = constant will form contour lines of constant x^3 and the net will appear as a system of curvilinear squares when viewed along the direction of the x^3 axis. Radial nets can be produced by introducing singularities ('sources' and 'sinks') into the function F at masks. This approach has the problem that it produces infinite values of x^3 at the singularities. This can be avoided by varying the tension coefficients in the cables so that the tension coefficients, T^α, in the direction of increasing Θ^α are given by

$$T^1 = \frac{f(\Theta^1)}{g(\Theta^2)}, \quad T^2 = \frac{g(\Theta^2)}{f(\Theta^1)} \tag{8.3}$$

The equilibrium equation therefore becomes

$$\left[\frac{f(\Theta^1)}{g(\Theta^2)}\, x^k_{,1}\right]_{,1} + \left[\frac{g(\Theta^2)}{f(\Theta^1)}\, x^k_{,2}\right]_{,2} = 0 \tag{8.4}$$

again for an unloaded net.

Introducing the new variables β^1 and β^2 such that

$$f(\Theta^1) = \frac{d\Theta^1}{d\beta^1}$$
$$g(\Theta^2) = \frac{d\Theta^2}{d\beta^2} \tag{8.5}$$

results in

$$\frac{\partial^2 x^k}{\partial \beta^{1\,2}} + \frac{\partial^2 x^k}{\partial \beta^{2\,2}} = 0. \tag{8.6}$$

We can therefore write

$$x^1 + i\,x^2 = F(\beta^1 + i\beta^2) \tag{8.7}$$

but still position the cables at constant spacing in Θ^1 and Θ^2.

Suitable choice of $f(\Theta^1)$ (or $g(\Theta^2)$ as appropriate) can ensure that the radial cables have finite length as the mast is approached when viewed in the direction of the x^3 axis. When viewed in this direction the net will form a system of curvilinear rectangles rather than curvilinear squares.

x^3 can be chosen to equal $c\beta^1$ (where c is a constant), except in the region of the mast. Here the continuum approximation will now break down and the x^3 coordinates can be found by solving a set of simultaneous equations which will give finite values of x^3. Figure 9 shows a drawing of such a net.

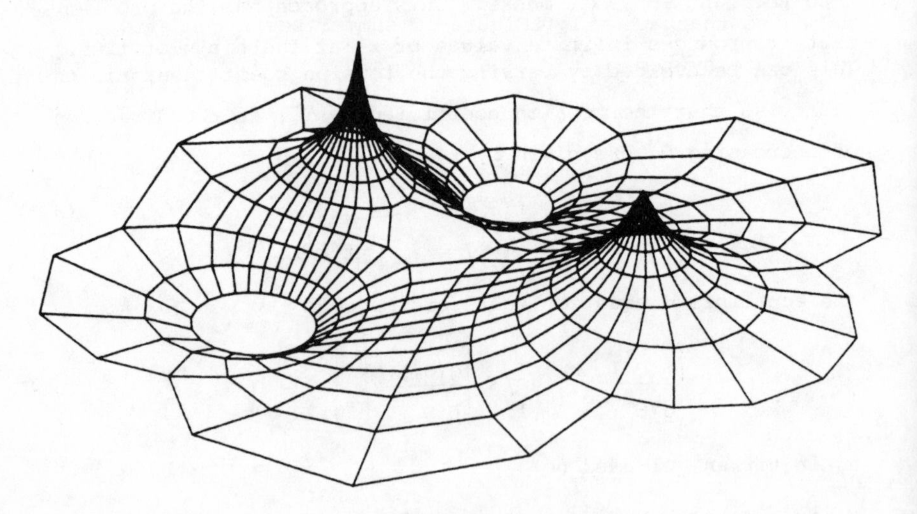

Figure 9

9. CUTTING PATTERNS FOR FABRIC STRUCTURES

The fabric for structures such as tents and air supported structures is woven in widths of, perhaps, 2m and strips of fabric are sewn, welded (by melting the coating) or glued together to form the surface. For many structures the Gaussian curvature of the surface is sufficiently high to require the strips to be shaped before being sewn together. The distortion of the fabric in forming a doubly curved surface is minimised if the cutting pattern is arranged so that the centre line of each panel is a geodesic on the surface. A geodesic is a line of zero geodesic curvature which is the curvature of a line on a surface seen when viewed along the normal to the surface.

If the coordinates on the surface are chosen to follow the direction of the yarns and the fabric strips are not too wide, the coordinates will approximate to a geodesic coordinate system on the surface, that is a set of geodesics and their orthogonal trajectories. Thus $a_{12} = 0$ and the condition that the lines θ^1 = constant are geodesics is

$$\mathbf{a}_{2,2}.\mathbf{a}_1 = a_{12,2} - \frac{1}{2} a_{22,1} = 0. \qquad (9.1)$$

Therefore a_{22} is a function of θ^2 only and can always be chosen so that $a_{22} = 1$. Figure 10 summarises a geodesic coordinate system.

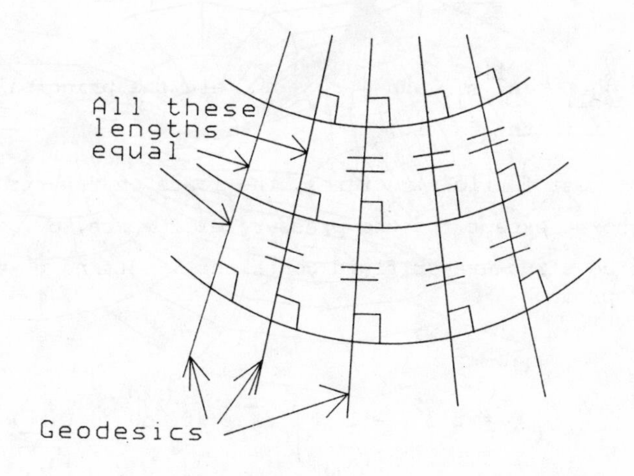

Figure 10

Substituting $a_{12} = 0$, $a_{22} = 1$ into Gauss' theorem, (2.23) gives

$$K = -\frac{1}{\sqrt{a_{11}}} (\sqrt{a_{11}})_{,22} \tag{9.2}$$

which shows how the width of the fabric strips is related to the Gaussian curvature of the surface. The finite widths of the real fabric strips means that there must be some relative rotation of the two sets of yarns away from their original orthogonal condition. This can be found using Gauss' theorem in the form (6.4) applied to equal mesh nets.

As stated previously, it is advisable to ensure that $n^{12} = 0$ under prestress conditions to avoid unpredictable elastic and creep strains in the bias direction of the fabric. Thus setting $a_{12} = n^{12} = 0$, $a_{22} = 1$ in the equilibrium equation (3.6) and scalar multiplying by $\mathbf{a}_1, \mathbf{a}_2$ and $\mathbf{a}_3$:

$$(\sigma_{(1)})_{,1} + p^1 a_{11} = 0$$

$$-\sigma_{(1)} (\sqrt{a_{11}})_{,2} + (\sigma_{(2)} \sqrt{a_{11}})_{,2} + p^2 \sqrt{a_{11}} = 0 \tag{9.3}$$

$$-\sigma_{(1)} \frac{b_{11}}{a_{11}} + \sigma_{(2)} b_{22} + p^3 = 0$$

where $\sigma_{(1)} = n^{11} a_{11}$ and $\sigma_{(2)} = n^{22}$ are the principal membrane stresses in the fabric.

For most fabric structures, the loads on the structure can be ignored except for the pressure difference, p^3, across an inflated structure or fluid container. Setting $p^1 = p^2 = 0$ in (9.3) produces

$$\sigma_{(1)} = f(\theta^2)$$

$$\sigma_{(2)} = f(\theta^2) - \frac{1}{\sqrt{a_{11}}} \int \sqrt{a_{11}} \frac{df}{d\theta^2} d\theta^2 + \frac{F(\theta^1)}{\sqrt{a_{11}}}. \tag{9.4}$$

Thus the surface (and the corresponding cutting pattern) are defined by the boundary conditions and the two functions of integration, $f(\theta^2)$ and $F(\theta^1)$ which can be chosen to try and obtain a surface to fulfil architectural or other criteria. Setting $f(\theta^2)$ = constant, $F(\theta^1) = 0$ corresponds to uniform surface tension as in a soap film and setting $f(\theta^2)$ = constant, $F(\theta^1) \neq 0$ corresponds to a soap film containing a set of tensioned threads which will automatically form geodesics on the surface.

A relatively simple numerical method for finding a surface and cutting pattern given suitable boundary conditions, $f(\theta^2)$ and $F(\theta^1)$ is described in two previous papers by the author [6], [8]. Figure 11 is taken from these papers and the 'simple' example of an air supported structure on a rectangular base was chosen because of the difficulty of producing a surface on a boundary with sharp corners. If the surface is not to be in the plane of the boundary at the corner, it is necessary for one of the principal membrane stresses to tend to zero at the corner.

Figure 11

CONCLUSIONS

Even though most tension structures are designed by numerical models which do not use the methods of differential geometry, there are occasions when the use of geometric ideas can produce interesting and useful forms. There are many ways of determining surfaces suitable for tension structures other than those considered in this paper. For example, one could consider an elastic membrane stretched over masts which could then be replaced by an orthogonal cable net in equilibrium following the directions of the principal stresses in the membrane. The elastic properties could be linear or non-linear and chosen to obtain a suitable surface.

REFERENCES

[1] Barnes, M.R., Review of Solution Methods for Static and Dynamic Analysis of Tension Structures. In *The Design of Air-Supported Structures*, pp. 133-146, The Institution of Structural Engineers, London, 1984.

[2] Barnes, M.R. and Wakefield, D.S., Dynamic Relaxation applied to Interactive Form Finding and Analysis of Air-Supported Structures. In *The Design of Air-Supported Structures*, pp. 147-161, The Institution of Structural Engineers, London, 1984.

[3] Green, A.E. and Zerna, W., *Theoretical Elasticity, 2nd Edition*, Oxford University Press, 1968.

[4] Struik, D.J., *Lectures on Classical Differential Geometry*, 2nd Edition, Addison-Wesley, Reading, Massachusetts, 1961.

[5] Tschebycheff, *Sur la coupe des vêtements, Oeuvres II*, p. 708, 1878.

[6] Williams, C.J.K., Form Finding and Cutting Patterns for Air-Supported Structures. In *Air-Supported Structures: The State of the Art*, pp. 99-120, The Institution of Structural Engineers, London, 1980.

[7] Williams, C.J.K. and Gaafar, I., Tear Propagation in Coated Fabrics, In *The Design of Air-Supported Structures*, pp. 73-83, The Institution of Structural Engineers, London, 1984.

[8] Williams, C.J.K., Form Finding and Cutting Patterns for Air-Supported Structures, In *The Design of Air-Supported Structures*, pp. 96-104, The Institution of Structural Engineers, London, 1984.

[9] Williams, C.J.K., The Integration of Wind Tunnel Testing and Structural Analysis of Air-Supported Structures. In *The Design of Air-Supported Structures*, pp. 109-113, The Institution of Structural Engineers, London, 1984.

[10] Williams, C.J.K., Methods of including the Effect of the Surrounding Air in the Dynamic Analysis of Air-Supported Structures. In *The Design of Air-Supported Structures*, pp. 175-189, The Institution of Structural Engineers, London, 1984.

GAUSSIAN CURVATURE AND SHELL STRUCTURES

C.R. Calladine
(Department of Engineering, University of Cambridge)

1. INTRODUCTION

It is widely agreed that Gauss's famous papers of 1825 and 1827 [4] on the curvature of general surfaces describe a very remarkable piece of work. But, paradoxically, few books on Differential Geometry succeed in explaining the essence of Gauss's ideas before they plunge into the complications of curvilinear coordinates, etc..

The aim of this talk is to explain the crucial features of Gauss's work in simple, elementary terms; to show that this simple treatment is directly useful in some purely geometrical applications; and to point out that it also holds an important key to an understanding of the mechanics of thin-shell structures. I shall follow the lucid exposition of Hilbert and Cohn-Vossen [5]; then I shall describe a scheme for thinking about and calculating a 'triangulated' version of a general surface; and finally I shall discuss briefly some topics which are treated at length in my book (Calladine [1]).

2. GAUSSIAN CURVATURE

Gauss introduced his ideas on the curvature of surfaces in his 1825 paper [4, p.81] by using an analogy with the curvature of a plane curve. The tangents at two points on a plane curve (see Fig. 1a) define the angle of embrace of a segment of a curve; and it is natural to define the curvature of the segment by the equation

$$\text{curvature} = \frac{\text{angle of embrace}}{\text{length of arc}} . \tag{1}$$

This is satisfactory for an arc of uniform curvature (the arc of a circle), but for the wider purpose of defining the curvature of a general smooth curve at a point, we may use the limit of (1) as the arc length of the segment tends to zero.

If (1) is taken as the definition of curvature, it is easy to show by means of elementary geometry that

$$\text{curvature} = (\text{radius of curvature})^{-1}, \tag{2}$$

where the radius of curvature is the radius of the circle that 'osculates' the curve at the point in question.

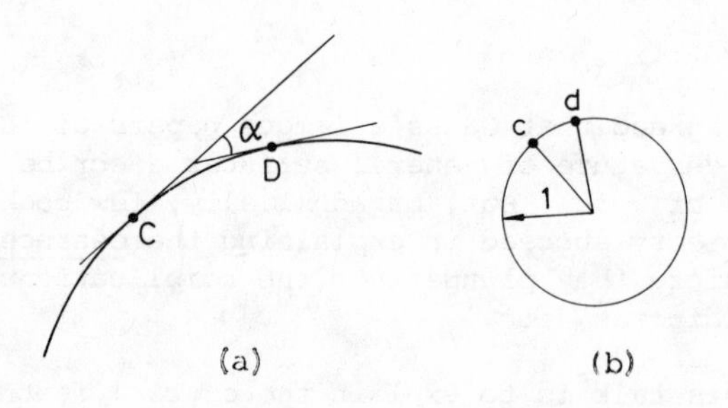

Fig. 1. (a) The segment CD of a plane curve subtends an 'angle of embrace' α. (b) Points C,D on the curve map to points c,d on the auxiliary circle of unit radius, by a rule of parallel normals.

At this point Gauss introduces the idea of an <u>auxiliary circle</u>, of unit radius, onto which points on a curve may be mapped by means of a rule of parallel normals. Thus, in Fig. 1, points C, D on the curve map to c, d, respectively, on the auxiliary circle. It is obvious that with this rule of mapping the <u>angle of embrace</u> of the segment CD is equal to the arc-length cd of the 'circular image' on the auxiliary circle; and thus that curvature may be defined, equivalently, by the appropriate limit of

$$\text{curvature} = \frac{\text{arc length cd of circular image}}{\text{arc length CD of curve}} . \tag{3}$$

Gauss saw that similar ideas could be applied to general surfaces in three-dimensional Euclidean space. Points on a surface may be mapped by a rule of parallel normals to points on an auxiliary sphere of unit radius [4, pp 3, 87] and so the closed boundary-curve defining a region of a surface has a 'spherical image' in the form of a closed curve on the auxiliary sphere: see Fig.2. Gauss defined the area of such a curve on the auxiliary sphere as the 'entire curvature' of the region on the original surface, by direct analogy with the plane curve and its 'circular image'.

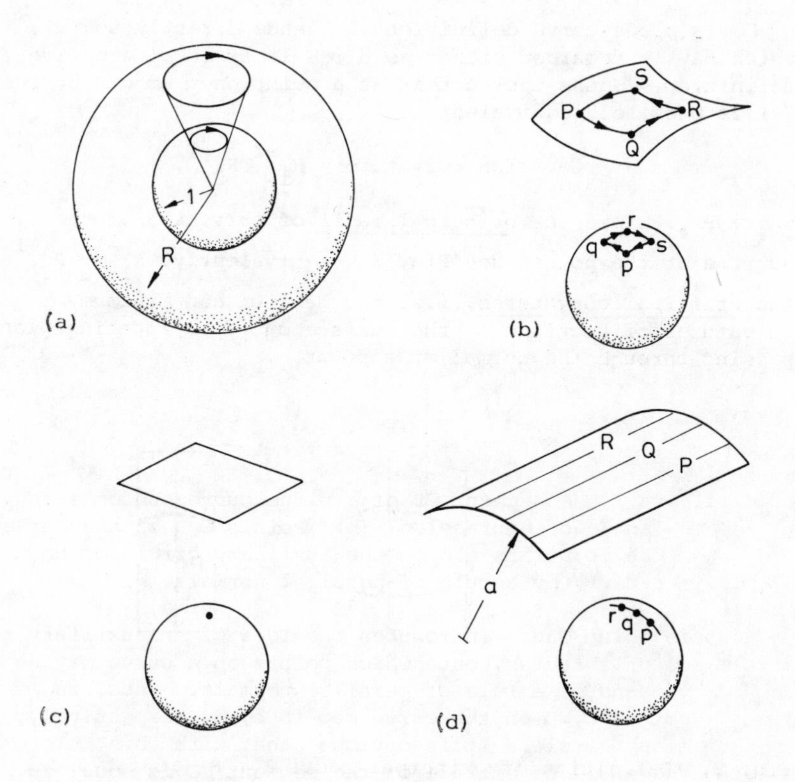

Fig. 2. Examples of the projection of portions of curved surfaces onto the unit sphere: (a) spherical surface, (b) surface having negative Gaussian curvature, (c) plane, (d) cylindrical surface.

The area of the spherical image is the 'solid angle of embrace' of the region of the surface. He then suggested, by analogy with (1), a definition of curvature of a surface:

$$\text{Gaussian curvature} = \frac{\text{area of spherical image}}{\text{area of region of surface}} . \qquad (4)$$

'Gaussian curvature' here replaces Gauss's own expression 'measure of curvature'. We shall use this term in relation to surfaces, leaving 'curvature' to refer to plane curves including - as we shall see later - <u>sections</u> of surfaces cut by planes through the normal at a point.

For a plane curve definition (1) leads directly to (2), which may be regarded either as a result or an alternative definition. Gauss showed that at a point on a smooth surface (4) is precisely equivalent to

$$\text{Gaussian curvature} = R_1^{-1} R_2^{-1} , \qquad (5)$$

where R_1, R_2 are the <u>principal radii</u> of curvature of the surface at the point: see Fig. 3. Equivalently, R_1^{-1}, R_2^{-1} are the <u>principal curvatures</u>, i.e. the maximum and minimum curvatures of sections of the surface cut by a rotating plane passing through the normal at a point.

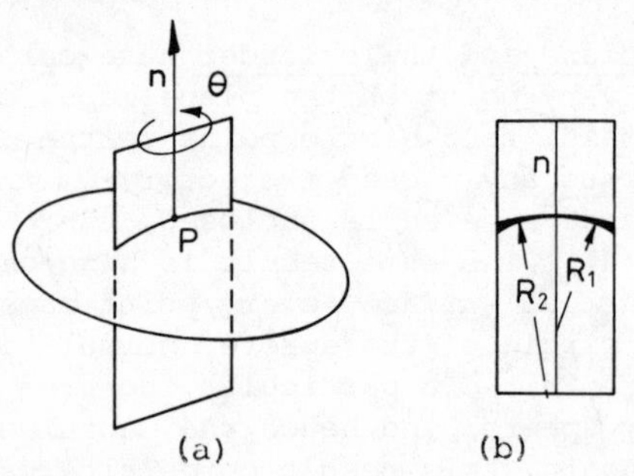

Fig. 3. Definition of principal radii R_1, R_2 at a typical point P on a smooth surface. (a) n is normal to the surface at P, and about it rotates a small plane. (b) During this process the curve of intersection varies between extremes having local radii of curvature R_1, R_2. (The corresponding 'principal' positions of the plane are separated by $\theta = \pi/2$).

The proof of this result is straightforward (e.g. Thomson & Tait [7, §138], Calladine [1, §5.5]) and we shall not repeat it here.

The simplest example of the precise equivalence of (4) and (5) is a sphere of radius R, as shown in Fig. 2a. In this case the linear dimensions of an arbitrary curve on the surface are R times those of the (exactly similar) image on the auxiliary sphere, so the ratio of areas is R^2; thus (4) leads to

$$\text{Gaussian curvature} = R^{-2}, \tag{6}$$

which agrees exactly with (5). For the sake of convenience the auxiliary sphere in Fig. 2a has been drawn concentric with the original sphere; but this is not necessary, of course.

In this example, in particular, a point on the surface and its image-point on the auxiliary sphere move in the same sense round their respective closed perimeters. For a saddle-shaped surface, on the other hand, (Fig 2b) whose principal curvatures are of opposite sign (since the respective centres of curvature lie on opposite sides of the surface), the Gaussian curvature is negative by (5); and this corresponds directly to the fact that the point on the surface and its spherical image move in opposite senses round their respective closed curves, as shown.

Both the <u>plane</u> and the <u>cylinder</u> have zero Gaussian curvature. Every point on the plane maps, by the rule of parallel normals, to a single point on the auxiliary sphere (Fig. 2c). Thus any closed trajectory on the plane maps to a point; and since no area is enclosed, the Gaussian curvature is zero, by (4). The same result is also obvious from (5). With a cylindrical surface, every point maps to a point on a single great circle of the sphere, normal to the axis of the cylinder (Fig. 2d). In particular, no area can be enclosed on the auxiliary sphere; and hence the Gaussian curvature is here also zero. Again, this result comes alternatively from (5), since one principal curvature is zero.

In all these examples we see that the two definitions (4) and (5) lead to identical results; as indeed they must in general, of course, in accordance with Gauss's result.

At this point it is tempting to argue that, since the two definitions are precisely equivalent, one of them is unnecessary; and thus that we should decide which one of the two is the more convenient for general purposes, and then discard the other.

The main point of the present paper is that in general it is best to retain both definitions of Gaussian curvature, since they deal with two distinct, but equally important, aspects of the geometry of surfaces. As we shall see, the use of both definitions simultaneously constitutes a powerful tool for solving some geometrical problems; and by this means we can also perceive more clearly the mode of mechanical action of thin shell structures. These topics will form the subject of Sections 5 and 6 respectively. But we are not ready for these examples until we have established one further result, in Section 4. It concerns the details of the physical interpretation of definition (4).

3. SURFACES IN THREE AND TWO DIMENSIONS

When we move on from simple examples such as cylinders and spheres to general surfaces, we face in particular an awkward problem over the specification of a coordinate system. On the surface of a cyclinder we can use a cartesian system of two sets of parallel coordinate lines which intersect each other orthogonally, and on the surface of a sphere we can use two sets of orthogonal lines in the form of circles of latitude and longitude. These examples illustrate the fact that two coordinates are needed to locate a point on a surface. In this sense surfaces are two-dimensional. They are, of course, also three-dimensional in the sense that they exist in three-dimensional Euclidean space. The two definitions (4), (5) of Gaussian curvature correspond exactly to these two different views of surfaces. It is obvious, of course, that (5) is a three-dimensional definition, since the idea of principal curvatures is essentially three-dimensional: it requires (see Fig. 3) a normal to the surface about which to rotate a local 'cutting plane'. It is not so obvious, however, that definition (4) is two-dimensional; and indeed, it appears at first sight that it is also three-dimensional, since it uses a solid auxiliary sphere (see Fig. 2). In the next section we shall demonstrate that (4) is in fact a statement about surfaces conceived as being two-dimensional.

We close this section with the remark that one set of orthogonal coordinate lines on the surface of the sphere - the lines of longitude - are not parallel to each other. Thus the distance measured on the surface between two points on a given latitude circle depends not only on the longitudes of the two points but also on the latitude. This is obvious, of course. But for the present purposes we note in particular the contrast between this case and that of a plane cartesian coordinate system: in cartesian coordinates the distance between two points on a line of y = constant depends only on the x-coordinates of the two points, and <u>not</u> on the y-coordinate. This example demonstrates that the two-dimensional surface of a sphere is <u>non-Euclidean</u>. Features of Euclidean plane geometry - e.g. that the interior angles of a triangle sum to π(radians) - do not hold on the surface of a sphere, or indeed on a general curved surface. As we shall see, these non-Euclidean features of two-dimensional surface geometry are a direct consequence of the non-zero Gaussian curvature. Only if the Gaussian curvature is zero, as in the cylindrical surface, are these features of plane geometry preserved.

4. GEOMETRY OF TRIANGULATED SURFACES

The usual text-book treatment of the geometry of surfaces starts with a discussion of the nature of two-dimensional curvilinear coordinates. The working soon becomes relatively heavy and we quickly lose sight of the simple two-fold definition (4), (5) of Gaussian curvature. We can avoid the need for coordinate systems at this early stage by imagining that the general curved surface in question is replaced by a network of small plane triangular facets. Thus, for example, a sphere may be triangulated as shown in Fig. 4.

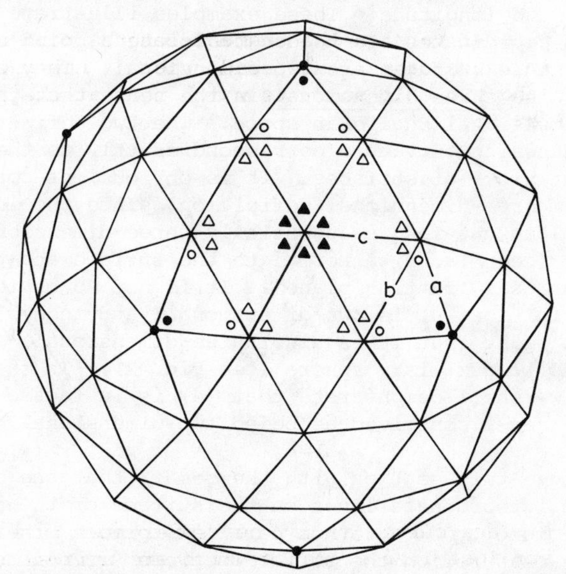

Fig. 4. A 'triangulated' version of a sphere, consisting of plane facets. The vertices marked ● correspond to the 12 vertices of the primitive regular icosahedron, each of whose 20 faces has been subdivided into 9 small triangles. The pattern of subdivision involves two kinds of isosceles triangle, having edges ab and bc, respectively. Angles and lengths are computed in the text.

Some features of a surface are obviously lost in such a process, while others are preserved. The situation is analogous to that by which points on a plane curve may be marked out by means of a sequence of straight-line segments set off at a small angle α from each other, as shown in Fig. 5. If the lengths of all segments are equal to ℓ, and the angles are all equal, as in Fig. 5b, then the points lie on a circle whose curvature is given, approximately, by

$$\text{curvature} = \alpha/\ell. \tag{7}$$

This is the appropriate version of (1), and α is measured in radians. The radius of the circle is, of course, actually equal to $\frac{1}{2}\ell/\sin\frac{1}{2}\alpha$; and so (7) is accurate to the extent that the approximation $\sin(\frac{1}{2}\alpha) \approx \frac{1}{2}\alpha$ is satisfactory. Thus, for example, if $|\alpha| < 9^{\circ}$, the error in radius is smaller than 0.1%. The same scheme may be used to set out curves whose curvature varies with arc-length. It is simplest to use segments of constant length ℓ; then at each kink α is chosen so that

$$\alpha = \text{curvature} \times \ell. \tag{8}$$

In a more general version the segment-lengths need not be constant; in which case (7) is used, where ℓ now represents the sum of the two half-segments which meet at the angle α, as shown in Fig. 5c.

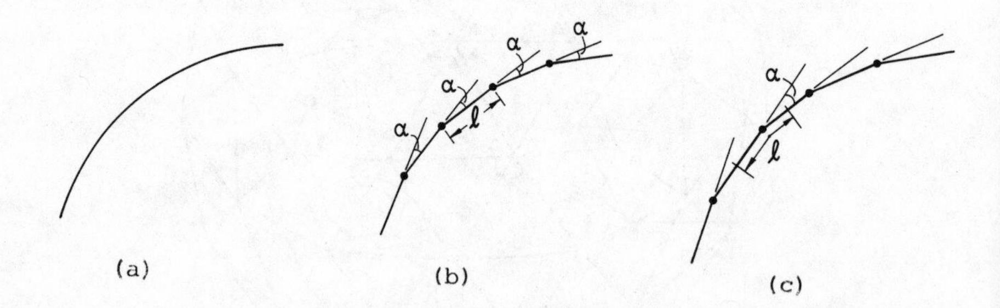

Fig. 5. A plane curve (a) may be represented by a sequence of straight-line segments which are uniform (b) for a circle, but non-uniform (c) in general. In (c) each vertex is associated with the sum of half-lengths of the segments concerned.

In all such schemes any inaccuracy in the position of the points in relation to the intended curve is associated with the small difference between the contour length of the intended curve and the segmented version. This is negligible provided the values of α are suitably restricted, as above.

What is the corresponding rule for the construction of a surface of given Gaussian curvature? It is, simply:

$$\text{Gaussian curvature} = \frac{\text{angular defect at a vertex}}{\text{area associated with the vertex}} \quad . \tag{9}$$

Here, the 'angular defect at a vertex' (Fig. 6a) is defined as

2π - sum of interior angles of faces meeting at the vertex, (10)

while the 'area associated with a vertex' is equal to 1/3 of the areas of the triangles meeting at the vertex.

This simple rule follows directly from definition (4) as soon as two geometrical propositions, as follows, are understood.

(1) It is only the vertices of the figure (and not the faces or the edges) that subtend solid angle.

(2) The solid angle of embrace of a vertex is exactly equal to the angular defect of the vertex (Fig. 6).

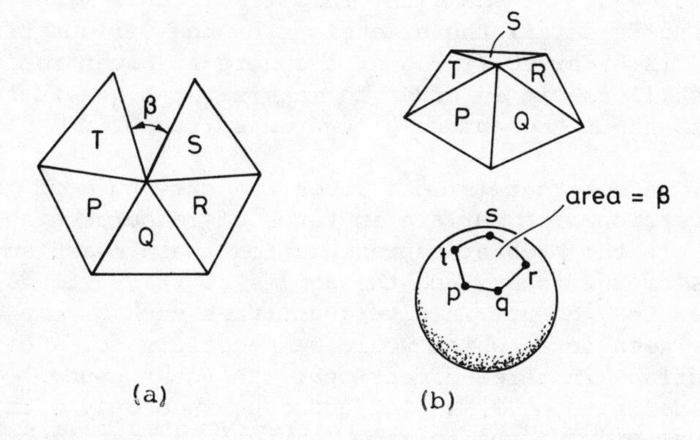

Fig. 6. A typical vertex of a faceted surface (a) has 'angular defect' = β when cut and flattened; and (b) has 'solid angle of embrace' (= area enclosed on the unit sphere) = β.

Proposition (1) is easy to prove. It may be shown by inspection that the area of the spherical image of a simply closed path on the surface of a polyhedron is zero when the path lies entirely on one face, or when it passes from one face to a neighbouring face (and back) without encircling a vertex; or indeed when it lies on any number of faces provided only that it does not enclose any vertices. The only closed paths on the polyhedron whose spherical images do enclose area on the auxiliary sphere are those which enclose vertices.

Proposition (2) appears to have been published first as Question vi in the Cambridge University Mathematical Tripos examination on the afternoon of Monday 18 January 1869. The question was probably set by J.C. Maxwell.

The proof, given in [5, §29], [1, §5.5.1], depends on a classical Greek theorem which states that the area of a region of a unit sphere bounded by arcs of great circles depends only on the sum of the exterior angles of the figure (for a proof see, e.g., [1, Appendix 7]). Perhaps the most remarkable feature of this result is that the solid angle subtended by an isolated vertex is unchanged when the vertex (having at least four facets) is 'folded' arbitrarily by having the 'crease' or 'dihedral' angles between the facets changed: the solid angle subtended at the vertex is independent of these dihedral angles.

In Section 5.2 we shall use this result as a means of calculating in detail the dimensions of the various triangles in Fig. 4 in order to represent a sphere of given radius; but here we shall use it in order to make several qualitative statements about the nature of curved surfaces.

First, we see that (9) describes the curvature of the faceted version of a surface in terms of measurements on the surface. In the present representation these measurements are the areas of the facets and the angles of the vertices; and in particular they do not involve quantities such as the dihedral angles between faces which would be necessary for a description of the surface in three-dimensional Euclidean space. We thus see that definition (9) is in fact concerned only with a two-dimensional view of the surface. This is a most important result; and the present derivation of it seems to be the simplest.

Second, we can prove immediately the well-known theorem that inextensional deformation of a surface preserves the Gaussian curvature. An inextensional deformation is defined as one in which all lengths within the surface are unchanged; and so in the faceted representation the sides and angles of the facets do not change. It is clear from (9) that in these circumstances the Gaussian curvature does not change. We are supposing here, of course, that the surface does actually deform in three-dimensional Euclidean space in the sense that at least some of the dihedral angles between the facets change. This will not normally be possible if the surface is complete and simply-connected: in order to realise an inextensional bending deformation it is usually necessary to 'cut out' a portion of the surface. This type of inextensional bending deformation is well illustrated by the way in which a piece of orange-peel covering, say, 1/3 of the area of the orange, can be 'rolled up' into a cigar-shaped object. In this operation one principal radius at each point increases and the other decreases, so that the product remains invariant (see (5)).

Third, it is worth noting that although we are here concerned with the use of small angular defects in the approximate representation of smooth surfaces by faceted ones, proposition (2) applies equally to large angular defects. Thus, for example, the angular defect at a vertex of a cube is $\pi/2$. Consequently each vertex subtends a solid angle of $\pi/2$, and so the eight vertices together subtend a solid angle of 4π. This is not surprising, since the total solid angle enclosed by the vertices of any simply closed polyhedron is 4π; and indeed it is obvious that the sum of the angular defects over all the vertices of any such figure is 4π. This result may also be derived independently from the famous topological theorem of Euler, which states that

$$f - e + v = 2, \tag{11}$$

where f, e, v are the number of faces, edges and vertices, respectively. The 2 on the RHS of (11) corresponds to the total solid angle of 4π.

5. EXAMPLES OF THE USE OF GAUSSIAN CURVATURE

Here we describe two examples which illustrate the way in which an understanding of the two-fold nature of Gaussian curvature leads to very simple calculations.

5.1 Example 1

The 'formwork' for a shallow doubly-curved reinforced-concrete shell roof consists of a curved timber grid-work onto which almost-rectangular plywood panels are to be nailed. How should the panels be cut, given that the principal radii of curvature of the uniformly curved surface are R_1, R_2?

Suppose we plan to use panels of approximate size $a \times b$ (where $a, b << R_1, R_2$), with four corners meeting at a vertex.

Associated with each vertex will be an area ab, one-quarter of which comes from each of four panels. It follows immediately from (9) that the required (small) angular defect, δ (radians) at each vertex is given by

$$\delta = ab/R_1R_2. \tag{12}$$

If the panels in the central 'cross' are rectangular, the other panels should be cut with shear angles of $\pm\delta$, $\pm 2\delta$, etc., in accordance with the scheme shown in Fig. 7; it is easy to verify that this scheme gives an angular defect of δ at each vertex.

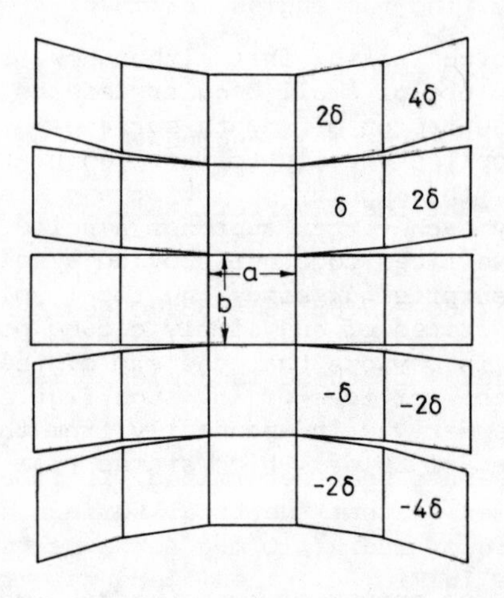

Fig. 7. Cutting-pattern for near-rectangular parallelograms to fit on a shallow surface having principal radii R_1, R_2. The 'shear' angles marked give an angular defect of δ at each vertex, where $\delta = ab/R_1R_2$.

5.2 Example 2

A geodesic dome is to be built with the layout shown in Fig. 4, in which each of the 20 triangles of the primitive icosahedron is sub-divided into 9 triangles (i.e. T = 9 in the notation of Caspar & Klug [2] and Coxeter [3]). The radius of the dome is to be 10 m. Specify the dimensions of the triangles.

The chosen pattern of triangulation of the sphere has $20 \times 9 = 180$ faces and 92 vertices. Suppose as a first approximation that an equal area is associated with each vertex. Then the required angular defect per vertex is

$$4\pi/92 = 0.1366 \text{ radian} \\ = 7.83 \text{ degrees} . \qquad (13)$$

Figure 4 shows a scheme in which there are two kinds of triangle. Five equal triangles with angles marked ● meet at a vertex, so

$$\bullet = (360 - 7.83)/5 = 70.43 \text{ degrees}. \qquad (14)$$

Since the sum of the interior angles of the triangle is 180°, we can immediately find the angles marked o:

$$o = (180 - 70.43)/2 = 54.78 \text{ degrees}. \qquad (15)$$

There are several ways of arranging the remaining 6 triangles. Perhaps the simplest is to make them all congruent, with equal angles ▲ at the 6-vertex: this angle is determined by

$$\blacktriangle = (360 - 7.83)/6 = 58.70 \text{ degrees},$$

and the other angles ▲ of these isosceles triangles follow directly.

Once the angles have been determined, it is easy to fix the lengths of the edges so that the total surface area is equal to that of a sphere of radius 10 m. Let a be the length of the edges marked a in Fig. 4. Then elementary trigonometry, using the angles already determined, gives the areas of the two kinds of triangle as $0.471\ a^2$ and $0.591\ a^2$, respectively. The total area of the triangles is given by

$$(60 \times 0.471 + 120 \times 0.591)a^2 . \qquad (16)$$

Assuming that the total area of the facets is negligibly different from that of the sphere, we equate this this to $400\pi\ m^2$, and hence obtain

$$a = 3.56 \text{ m}. \qquad (17)$$

It follows directly that the other lengths are given by

$$b = 4.10 \text{ m}, \\ c = 4.19 \text{ m}. \qquad (18)$$

This example illustrates the rapidity with which the calculations may be performed. It might be objected, of course, that if 1/3 of the area of each triangle is to be associated with each of its vertices, then the area associated with the three different kinds of vertex will be somewhat different, in contradiction to the initial assumption. It is easy enough to compute the area associated with the vertices of the polyhedron calculated above, and then to repeat the calculation for a more equable distribution of angular defect. The resulting figure is more closely spherical; but even if several iterations of this kind were needed, the calculations would still be nugatory in comparison with those which would have to be done by conventional 'solid' geometry (see, e.g. Kenner [6]).

6. MECHANICS OF THIN-SHELL STRUCTURES

The dual nature of surfaces has a direct connection with the mechanics of distortion of thin elastic shell structures under load. This is a large subject (see, e.g. Calladine [1]) and I shall give here only a brief sketch of the crucial part played in the theory by Gaussian curvature.

Consider a uniform thin-walled spherical shell with, say, thickness = radius/100. If it is loaded by interior pressure, it carries the uniform load by means of uniform 'membrane' tensile stress, in essentially the same way that a soap-bubble contains interior pressure; and the calculation of stress from the equations of equilibrium of an element is thus particularly straightforward. But if the same shell is now loaded instead by diametrically opposed <u>point</u> forces, as in Fig. 8, the behaviour is much more complicated, because now there will be a small region of the shell in the vicinity of each load in which there exist considerable <u>bending stresses</u>.

The simplest example of bending stresses in a structure is a beam under a transverse force, as shown in Fig. 9. The beam, being straight, cannot deploy a state of tension in order to equilibrate the applied load. Flat plates carry applied transverse forces by a sort of two-way 'beam' action, which again does not involve tension in order to balance the applied load. We are assuming here that the beam is made of a stiff material, and that its deflection is small compared to its span.

Now a study of the equations of thin-shell structures reveals that in general shells carry the forces which are applied to them by a combination of 'membrane' and 'bending' effects. In some regions of a shell one mode of action will be dominant, while in the remaining parts the other mode will be more important.

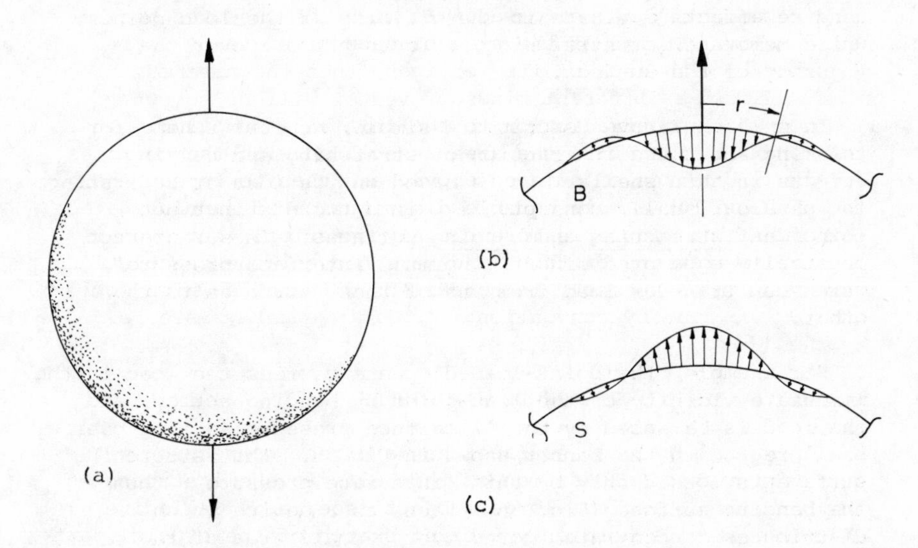

Fig. 8. (a) A uniform thin-walled spherical shell, made from (stiff) elastic material, is subjected to diametrically opposed 'point' loads. (b), (c) The shell is split, conceptually, into separate bending (B) and stretching (S) surfaces, which are endowed with the corresponding mechanical properties of the shell. See section 6 for details.

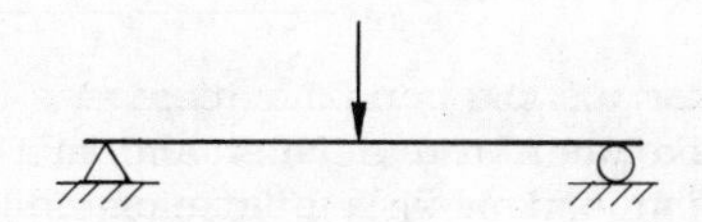

Fig. 9. A straight elastic beam carries an applied transverse load by flexural effects.

In the spherical shell shown in Fig. 8a, for example, bending effects dominate in the vicinity of the load-points, while membrane, or stretching, effects prevail over the majority of the surface.

In my book I have described a simple, general scheme for thinking about the interaction of stretching and bending effects in thin shells. It is based on the idea of separating the physical shell, conceptually, into two distinct but coincident stretching and bending surfaces. These interact physically by a predominantly normal 'interface pressure', which can transfer load from one of these surfaces to the other.

For example, the point-loaded sphere carries the load in the immediate vicinity of the load-point by bending action, and the load is balanced by the 'interface pressure' acting over a small region in the manner shown in Fig. 8. The 'stretching' surface is loaded only by this 'interface pressure': thus the bending surface (Fig. 8b), acts as a device which distributes concentrated force over a small area of the stretching surface so that this surface is not called upon to sustain singularities of pressure.

It is difficult to describe interactive effects such as these without giving details of the equations which govern the behaviour of the thin shell or, equivalently, of its two constituent surfaces. It is clear that the preceding description describes the interaction from a purely statical point of view, involving the balance between the applied external forces and the internal stresses which are required by the laws of equilibrium of forces for a typical elementary part. The other ingredients of the problem, which I have not yet mentioned explicitly, are

(i) the elastic law of the stiff material, i.e. the general relationship between the stress and the strain or geometrical distortion which it causes; and

(ii) the purely kinematical relations between the (small) strains in the material and the (small) displacements of the shell surface.

In short, these further relations are concerned with the fact that the stresses - which are necessary for the statical equilibrium of every part of the shell - cannot occur without some small elastic distortion of the shell. In terms of the two-surface approach, the non-uniform membrane stresses in the stretching surface cause that surface to distort, while the bending stresses cause the bending surface to distort.

In the two-surface representation, the last condition to be stated is simply that the two surfaces must undergo equal geometric distortion, since they are physically coincident not only in the initial, stress-free, configuration but also throughout the history of deformation of the shell. It is clear that we must now consider the way in which each of the two surfaces distorts elastically.

Now in the preceding discussion of the geometry of surfaces we have been concerned particularly with the two distinct aspects of Gaussian curvature of an arbitrary given surface. It turns out that the key to the kinematical relationship between the distortion of the stretching and bending surfaces of the shell is the fact that the Gaussian curvature at corresponding points of the two surfaces has the same value, in both the original and the deformed configuration.

The <u>change</u> of Gaussian curvature in the stretching surface, with reference to the initial, stress-free state, is a consequence of straining of the surface. In terms of the faceted version of the surface the facets distort a little in their own planes, thereby changing slightly their vertex angles and hence altering the Gaussian curvature.

On the other hand, the change in Gaussian curvature of the bending surface comes about because the bending stresses <u>bend</u> the surface, thereby altering slightly the principal curvatures and thus the Gaussian curvature also.

This relationship between the change of Gaussian curvature in the two conceptual surfaces - calculated via a two-dimensional view (4) of stretching in the stretching surface and a three-dimensional view (5) of flexure in the bending surface - is the key kinematical connection between the two surfaces. It completes the set of governing equations, and thus enables problems to be solved.

This is not the place to describe the details of the calculations given in [1], which <u>inter alia</u> enable the precise distribution of interface pressure to be determined in a given case. But we may mention the fact that in the problem of the point-loaded spherical shell it turns out that the characteristic radius r of the zone of interface pressure (see Fig. 8b) is determined from the consideration that as r varies, other things being equal, the change of Gaussian curvature in the stretching and bending surfaces vary as r^{-2} and r^2 respectively; and so the two surfaces remain in contact with each other only when r has a particular value. The techniques of dimensional analysis, which are here applicable separately to the two surfaces without consideration in detail of the governing equations, show that

$$r \propto (Rt)^{\frac{1}{2}} , \tag{19}$$

where R, t are the radius and thickness, respectively, of the shell.

This simple and useful result emerges easily as a direct consequence of our understanding that the deformation of the two separate surfaces involves the distinct two- and three-dimensional views of surfaces which have formed the subject of this paper.

ACKNOWLEDGEMENT

I am grateful to Dr. A. Klug, of the M.R.C. Laboratory of Molecular Biology, Cambridge, and Peterhouse, for explaining Gaussian curvature to me.

REFERENCES

[1] Calladine, C.R., *Theory of shell structures,* Cambridge University Press, Cambridge, 1983.

[2] Caspar, D.L.D. & Klug, A., Physical principles in the construction of regular viruses, Cold Spring Harb. Symp. quant. Biol. **27**, pp 1-24, 1962.

[3] Coxeter, H.S.M., Virus macromolecules and geodesic domes, in *A spectrum of mathematics*, J.C. Butler (ed.), pp 98-107, Auckland University Press, Auckland, 1971.

[4] Gauss, K.F., *Disquisitiones Generales circa superficies curvas*, Göttingen, 1828 (English translation, *General investigation of curved surfaces* by J.C. Morehead & A.M. Hiltebeitel, Princeton, 1902; reprinted with introduction by R. Courant, Raven Press, Hewlett, New York, 1965).

[5] Hilbert, D. & Cohn-Vossen, S., *Geometry and the imagination,* translator P. Nemenyi, Chelsea Publishing Company, New York, 1952.

[6] Kenner, H., *Geodesic math and how to use it*, University of California Press, Berkeley, 1976.

[7] Thomson, W.T. & Tait, P.G., Treatise on natural philosophy, vol.1, part 1, Cambridge University Press, Cambridge, 1879.

MULTIVARIATE SPLINE ALGORITHMS

W. Boehm
(Technische Universitaet Braunschweig)

1. HISTORY

One of the simplest but most illustrative examples of multivariate splines are those over a regular triangular grid.

Triangular surface patches in CAGD were first considered by de Casteljau in 1963. He constructed his "surfaces a pôles" from a 3-D net of points (pôles) by repeated affine maps of a fixed triangle as shown in Fig. 1 (de Casteljau [6]). Later his "pôles" were named "Bezier points".

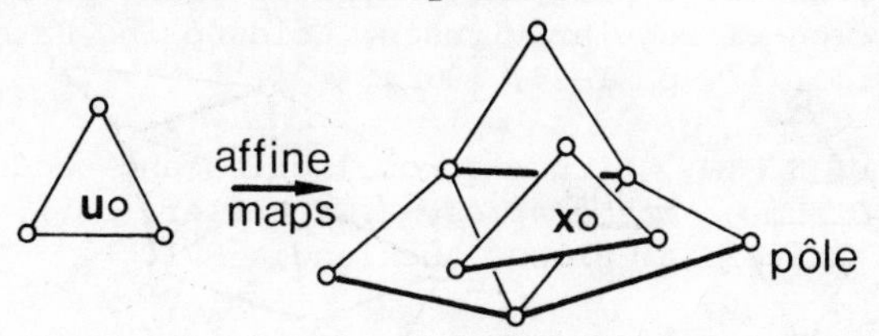

Fig. 1. Surfaces a pôles

Notation: We denote rows of coordinates, components or indices by lower-case bold letters. For simplicity, the d-dimensional real affine space A^d can be viewed as $\mathbb{R}^d$. An affine map is a mapping that leaves parallelism and linear interpolation unchanged. It should be noted that the given definitions and constructions are invariant under affine maps (see eg. Boehm et al. [3]).

B-splines over triangles were first developed by Sabin in 1977. He constructed them from a lower degree B-spline by the use of convolutions with the unit pulse (Sabin [12]). Fig. 2 shows an example: a quadratic B-spline is constructed from a linear B-spline. The continuity class in direction of the convolution is increased by one.

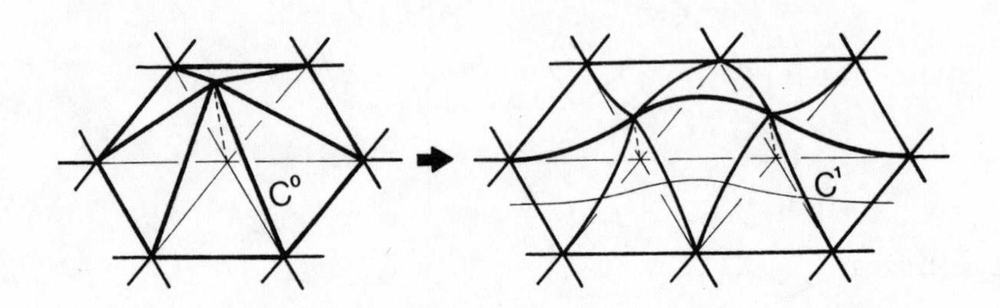

Fig. 2. Sabin's B-splines

General C^r conditions for adjacent triangular patches were considered by Farin in 1979: for example, two such patches are C^1 continuous if each pair of adjacent triangles of Bézier points is the affine image of the corresponding pair of triangles in the parameter plane, as shown in Fig. 3 (Farin [10] see also Sabin [12]).

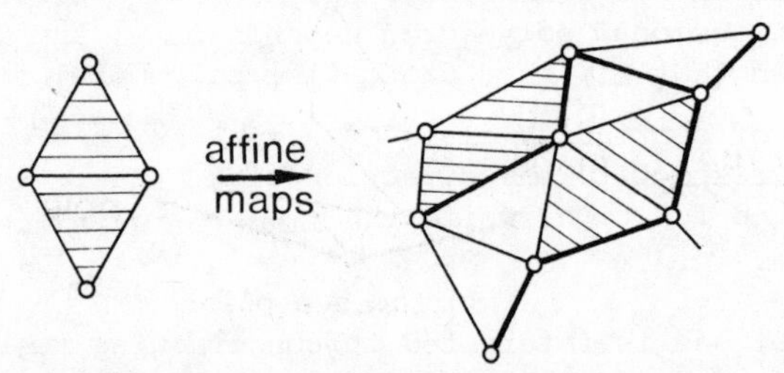

Fig. 3. Farin's C^r-condition, r=1

Multivariate splines were increasingly considered in the early eighties by de Boor, Hoellig and de Vore as well as by Dahmen and Micchelli. The theory of multivariate splines has undergone rapid development since this time (de Boor et al. [5], Dahmen et al. [8]).

2. MULTIVARIATE SPLINES

Multivariate spline algorithms are easily developed from a geometric definition of splines, given by Schoenberg in 1966 (see eg. Dahmen et al. [8]). A polyhedron P that spans the affine space A^m is affinely mapped into an affine space A^r, $r \leq m$, as in Fig. 4.

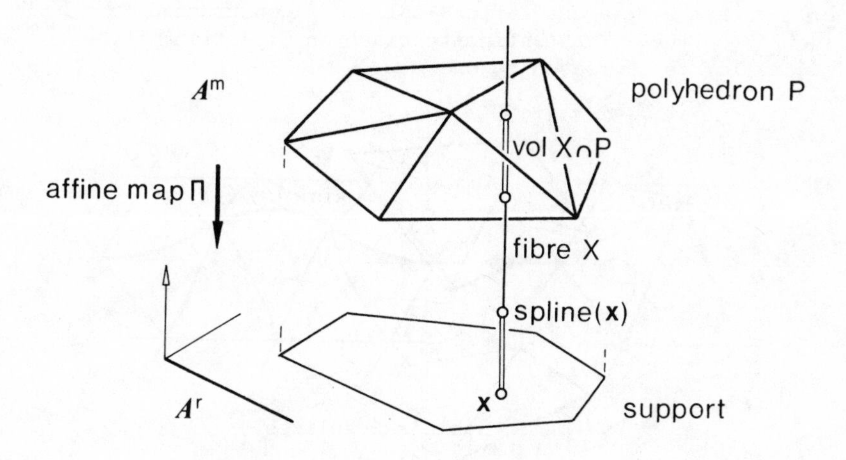

Fig. 4. Schoenberg's construction

All points $\mathbf{y}$ of A^m that have the same image $\mathbf{x}$ in A^r form the fibre X of $\mathbf{x}$. This fibre is an m-r dimensional affine subspace of A^m. The intersection of X with the polyhedron P, $X \cap P$, therefore forms an m-r dimensional polyhedron. The volume, V, of this m-r dimensional polyhedron depends on $\mathbf{x}$, and it can easily be shown that $V(\mathbf{x})$ is a piece-wise polynomial of degree $d=m-r$ and is of continuity class C^{d-1} if P is in a general position with respect to the direction of X. Hence the volume $V(\mathbf{x})$ is a spline function, while the image of P is the support of this spline.

In practice, one is interested in constructing implementable B-splines. They are developed from special polyhedrons: simplices and boxes (see e.g. Dahmen et al [8]).

3. SIMPLEX SPLINES

A simplex S in A^m is given by m+1 vertices $\mathbf{v}_0,\ldots,\mathbf{v}_m$ that span A^m. Schoenberg's construction gives a "simplex spline" where the images $\mathbf{x}_i$ of the vertices $\mathbf{v}_i$ are called the "knots". By assuming a unit volume for the simplex, one can normalize the simplex spline: it then depends on the knots $\mathbf{x}_i$, only, but not on the vertices $\mathbf{v}_i$ at the fibres $X(\mathbf{x}_i)$ of these knots.

We denote this normalized simplex spline by

$$M(\mathbf{x}) = M(\mathbf{x} \mid \mathbf{x}_0, \ldots, \mathbf{x}_m),$$

see Fig. 5.

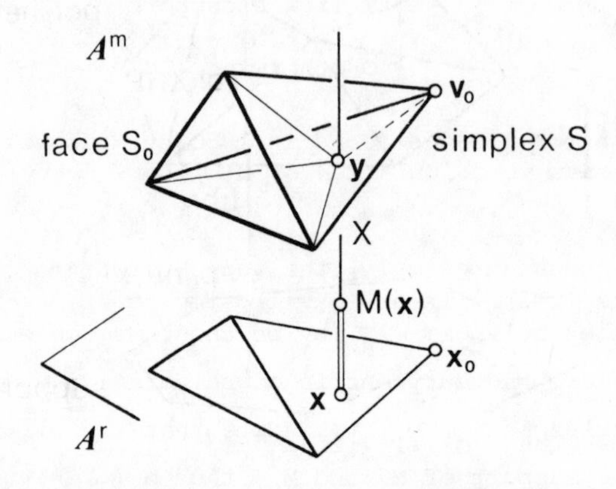

Fig. 5 Simplex spline

Any m vertices, say $\mathbf{v}_1, \ldots, \mathbf{v}_m$, form an m-1 dimensional simplex S_0 and a corresponding (normalized) simplex spline $M_0(\mathbf{x})$. It is easy to show that the value of $M(\mathbf{x})$ at $\mathbf{x}=\mathbf{x}_0$ can be expressed by the value of $M_0(\mathbf{x})$ at $\mathbf{x}=\mathbf{x}_0$:

$$M(\mathbf{x}_0) = \frac{m}{m-r} M_0(\mathbf{x}_0) \qquad (*)$$

To compute $M(\mathbf{x})$ at any given $\mathbf{x}$ one subdivides the simplex S into subsimplices formed by the "faces" S_i and a point

$$\mathbf{y} = \lambda_0 \mathbf{v}_0 + \ldots + \lambda_m \mathbf{v}_m$$

in the fibre of $\mathbf{x}$, where the λ_i are the barycentric coordinates of $\mathbf{y}$ with respect to the $\mathbf{v}_i$ (see Fig. 5). Using the well-known volume property of these coordinates together with their affine invariance, i.e.

$$\mathbf{x} = \lambda_0 \mathbf{x}_0 + \ldots + \lambda_m \mathbf{x}_m$$

one gets if $m \neq r$

$$M(\mathbf{x}) = \frac{m}{m-r} (\lambda_0 M_0 (\mathbf{x}) + \ldots + \lambda_m M_m (\mathbf{x})).$$

If m=r one has

$$M(\mathbf{x}) = \begin{cases} 1 \text{ if } \mathbf{x} \in \text{support } M \\ 0 \text{ otherwise} \end{cases}$$

This recursion formula was found in 1980 by Micchelli. It allows the recursive calculation of $M(\mathbf{x})$ for any $\mathbf{x}$ (Dahmen et al. [8]).

Figure 6 shows an example, the support for the composition of a bivariate (r=2), linear (m-r=1) simplex spline from four constant simplex splines: λ_3 may be chosen to be zero, so that λ_0, λ_1 and λ_2 are the barycentric coordinates of $\mathbf{x}$ with respect to the knot triangle $\mathbf{x}_0$, $\mathbf{x}_1$, $\mathbf{x}_2$. Note that the plotted $\mathbf{x}$ lies outside of the support of M_1 and M_2, thus $M_1(\mathbf{x}) = M_2(\mathbf{x}) = 0$ and $M(\mathbf{x}) = \frac{3}{1} \lambda_0 M_0 (\mathbf{x})$.

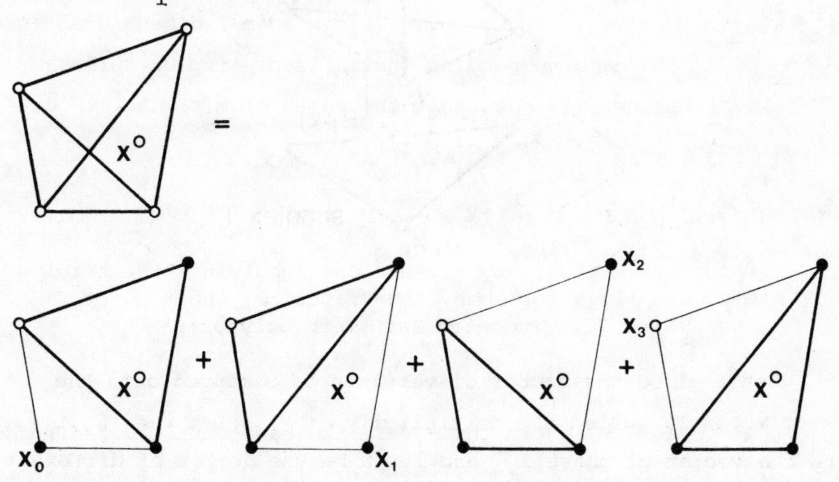

Fig. 6. Micchelli's recursion

The directional derivative of a function $F(\mathbf{x})$ at $\mathbf{x}$ in the direction $\mathbf{v}$ is defined by

$$D_{\mathbf{v}} F(\mathbf{x}) = \frac{d}{d\tau} F(\mathbf{x}+\tau\mathbf{v}) \Big|_{\tau=0}$$

If the function is a simplex spline $M(\mathbf{x})$ and $\mathbf{v}=\mathbf{z}-\mathbf{x}$ one gets

$$D_{\mathbf{v}}\ M(\mathbf{x}) = m \sum_{i=0}^{m} (\mu_i - \lambda_i)\ M_i(\mathbf{x})$$

where λ_i and μ_i are the barycentric coordinates of $\mathbf{x}$ and $\mathbf{z}$, respectively (see, eg. Dahmen et al. [8]).

4. TRUNCATED BERNSTEIN POLYNOMIALS

If more than one vertex of the simplex S are mapped into one knot, the simplex spline degenerates. We consider a very special case, in which the image of S is also a simplex in A^r, say T, see Fig. 7.

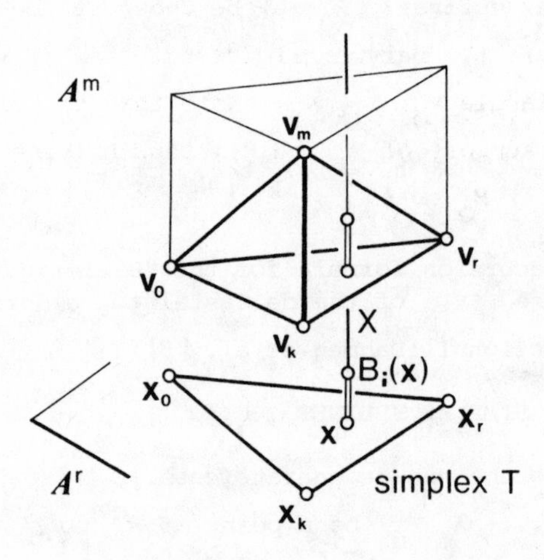

Fig. 7. Truncated Bernstein polynomial

Let i_k+1 be the number of vertices of S mapped onto the knot $\mathbf{x}_k$ of T, called the multiplicity of $\mathbf{x}_k$. Let $\mathbf{i} = [\ldots, i_k, \ldots]$ be the vector of these i_k, and let μ be the number of different simplices S mapped onto T, ie. the number of different $\mathbf{i}$'s. Then the μth part of the simplex spline $M(\mathbf{x})$ will degenerate to the truncated Bernstein polynomial

$$B_{\mathbf{i}}\ (\mathbf{x}) = \frac{1}{\mu}\ M(\mathbf{x}),$$

and Micchelli's recursion formula becomes the well-known recursion formula for Bernstein polynomials

$$B_{\mathbf{i}}(x) = \sum_k \lambda_k B_{\mathbf{i}-\mathbf{k}}(x)$$

where $\mathbf{k}=[\,..,1,..\,]$ is a vector that has a 1 only at the position k. Note that $B_{\mathbf{i}-\mathbf{k}}$ is of degree m-r-1 (Dahmen et al. [8]).

One can consider linear combinations of Bernstein polynomials over T

$$\mathbf{b}(x) = \sum_{\mathbf{i}} \mathbf{b}_{\mathbf{i}} B_{\mathbf{i}}(x), \quad |\mathbf{i}| = \sum_k i_k = m-r,$$

with coefficients $\mathbf{b}_{\mathbf{i}}$. Expressing the $B_{\mathbf{i}}$ by the recursion formula above one gets

$$\mathbf{b}(x) = \sum_{\mathbf{j}} \mathbf{b}^1_{\mathbf{j}} B_{\mathbf{j}}(x), \quad |\mathbf{j}| = m-r-1,$$

where

$$\mathbf{b}^1_{\mathbf{j}} = \sum_k \lambda_k \mathbf{b}_{\mathbf{j}+\mathbf{k}} .$$

This dual recursion formula for the "Bézier points" $\mathbf{b}_{\mathbf{i}}$ represents one step of the de Casteljau algorithm, mentioned in Section 1 (Dahmen et al. [8]).

5. SUBDIVIDING SIMPLEX SPLINES

We return to the non-degenerate case. Let $\hat{y} = \hat{\lambda}_0 v_0 + \ldots + \hat{\lambda}_m v_m$ be a point of A^m having the image $\hat{x}$. Let $\hat{S}_0$ be the subsimplex of S defined by the vertices $\hat{y}, v_1, \ldots, v_m$, and let $\hat{M}_0(x)$ be the corresponding simplex spline, etc. From $S = \sum_k \hat{S}_k$, the recursion

$$M(x) = \sum_k \hat{\lambda}_k \hat{M}_k(x)$$

follows immediately for this "insertion of a new knot" (Prautzsch [11] see also Boehm [2]).

One can repeat this knot insertion. After inserting $\hat{x}$ up to m-r-1 times, $M(\hat{x})$ is expressed in Bernstein polynomials over different supports but all having the value 1 at $\hat{x}$. Thus, one can use knot insertion to compute $M(\hat{x})$ (Boehm [1], Prautzsch

[11]). Fig. 8 shows one step of the knot insertion algorithm for the simplex spline of Fig. 6. Note that $\mu=3$ here.

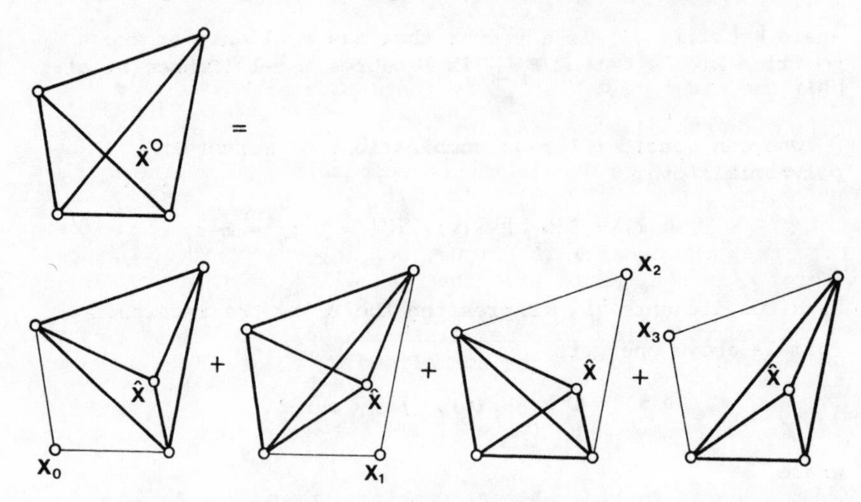

Fig. 8. Inserting a new knot

6. BOX SPLINES

A parallelpiped or box B of A^m is given by the point **b** together with m edges (vectors) $\mathbf{a}_1, .. \mathbf{a}_m$, forming an affine system of A^m, see Fig. 9. These edges are mapped onto a subset of the edges $\{\mathbf{u}_1, ... \mathbf{u}_n\}$ of a regular grid of points

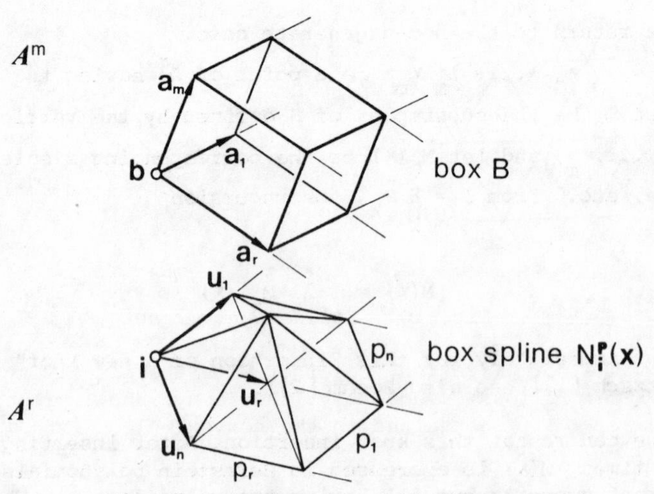

Fig. 9. Box spline definition

$$\mathbf{i} = \mathbf{o} + i_1 \mathbf{u}_1 + \ldots + i_n \mathbf{u}_n .$$

Then Schoenberg's construction gives a "box spline". Let $\mathbf{i}$ be the image of $\mathbf{b}$, let p_k be the number of box edges $\mathbf{a}_j$ that fall onto the edge $\mathbf{u}_k$, and let $\mathbf{p} = [\ldots, p_k, \ldots]$ be the vector of these numbers. We denote the corresponding box spline by $N_{\mathbf{i}}^{\mathbf{p}}(\mathbf{x})$, $\mathbf{p}$ is called the "order" and $\mathbf{i}$ the "origin" of $N_{\mathbf{i}}^{\mathbf{p}}$. Note that the order $\mathbf{p}$ is unique, but in general the position vector $\mathbf{i}$ is not. Note also that $|\mathbf{p}| = \sum_k p_k = m$.

In contradiction to general simplex splines, the box splines under consideration can be normalized by

$$\sum_{\mathbf{i}} N_{\mathbf{i}}^{\mathbf{p}}(\mathbf{x}) = 1, \text{ all } \mathbf{x} \in A^r .$$

Thus, one can consider linear combinations of box splines

$$\mathbf{s}(\mathbf{x}) = \sum_{\mathbf{i}} \mathbf{d}_{\mathbf{i}} N_{\mathbf{i}}^{\mathbf{p}}(x),$$

with coefficients $\mathbf{d}_{\mathbf{i}}$ forming a spatial grid that can be viewed as a sculptured image of the given grid in A^r, it is called the "control net" of the spline $\mathbf{s}(\mathbf{x})$ (see Fig. 10).

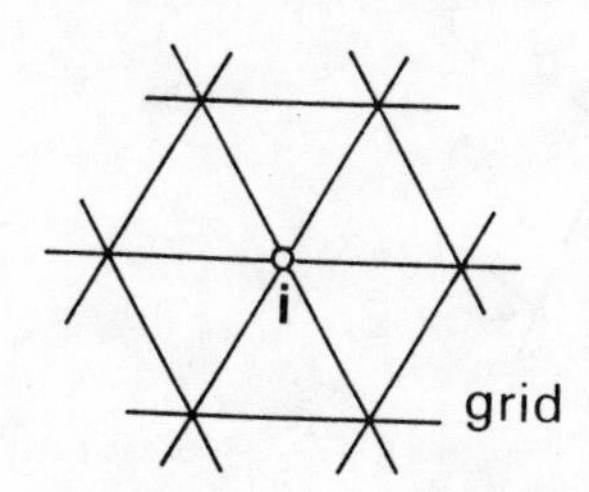

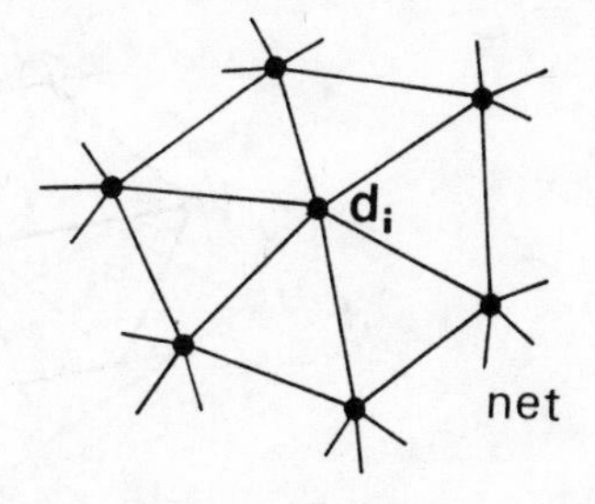

Fig. 10. Spline control net

From the normalization, together with the positivity of the box spline, follows immediately, the "convex hull property". Any point $\mathbf{s}(\mathbf{x})$ is contained in the convex hull of the $\mathbf{d}_{\mathbf{i}}$ which correspond to $N_{\mathbf{i}}^{\mathbf{p}}(\mathbf{x}) \neq 0$.

Note that any box spline $N_i^p(x)$ can be viewed as a "translate" of a primary box spline $N^p(x)$ that is

$$N_i^p(x) = N^p(x-i) ,$$

corresponding to the translate of the primary box B.

7. RECURSION OF BOX SPLINES

To compute $N_i^p(x)$ at any given x, we follow the same ideas as used in the case of simplex splines in Section 3. The faces of an A^m-box are A^{m-1}-boxes, generating box splines of order $p-k$, where k is as defined above. Let $x=i+\beta_1 u_1+\ldots+\beta_n u_n$ be a representation of x and let y be a point of the fibre X of x. One can consider the pyramids formed by the faces of B and the common vertex y (see Fig. 11).

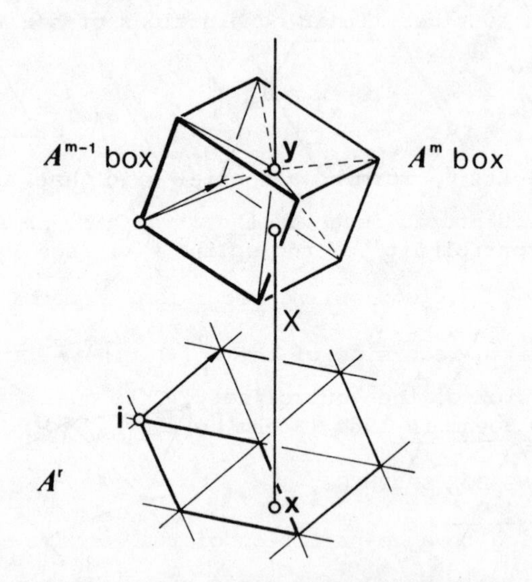

Fig. 11. Box spline recursion

Some geometric observations (Boehm [4]) yield

$$N_i^p(x) = \frac{1}{m-r} \sum_k (\beta_k \; N_i^{p-k}(x) + (p_k-\beta_k) \; N_{i+k}^{p-k}) ,$$

and if m=r (ie. $|\mathbf{p}|=r$)

$$N_{\mathbf{i}}^{\mathbf{p}}(\mathbf{x}) = \begin{cases} 1 \text{ if } \mathbf{x} \in \text{support } N_{\mathbf{i}}^{\mathbf{p}} \\ 0 \text{ otherwise} \end{cases}$$

This recursion formula was first found by de Boor in 1982 (de Boor et al. [5]). It allows the recursive calculation of $N_{\mathbf{i}}^{\mathbf{p}}(\mathbf{x})$ at any $\mathbf{x}$.

Figure 12 shows the supports for the composition of a bivariate (r=2) quartic (m-r=4) box spline, where $\mathbf{p}=[3,2,1]$. The $N_{\mathbf{i}}^{\mathbf{p}-\mathbf{k}}$ are solid lines, the $N_{\mathbf{i}+\mathbf{k}}^{\mathbf{p}-\mathbf{k}}$ are dashed and the needed factors are written inside the supports.

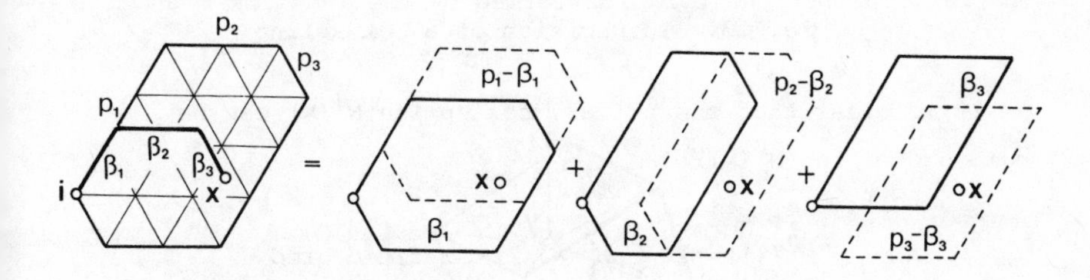

Fig. 12. Box spline recursion, example

With the given recursion of the $N_{\mathbf{i}}^{\mathbf{p}}(\mathbf{x})$, there exists a related recursion of the control net. This dual recursion was considered by Boehm in 1984 (Boehm [4]).

8. SUBDIVIDING BOX SPLINES

Corresponding to a p-partition of its edges, one can subdivide a primary box B of A^m into p^m identically shaped subboxes. This subdivision yields a subdivision grid of points $\mathbf{i}$ together with subbox splines $\hat{N}_{\mathbf{i}}^{\mathbf{p}}(\mathbf{x})$, as shown in Fig. 13.

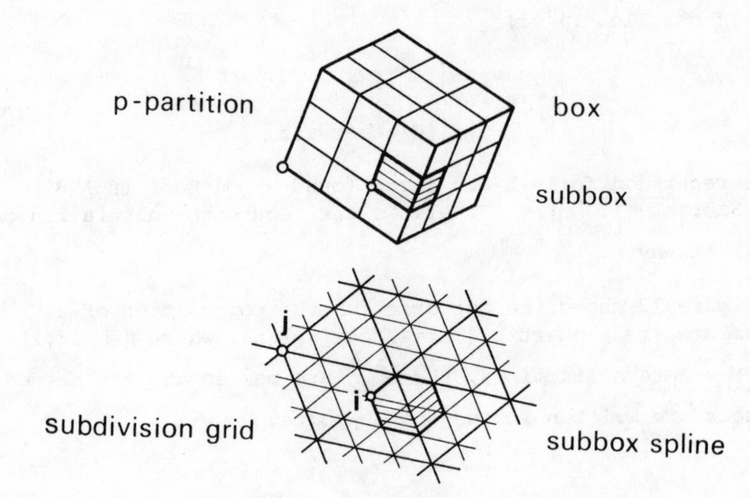

Fig. 13. Subdivision of a box spline

It is clear that the primary box spline $N^{\mathbf{p}}(\mathbf{x})$ can be composed of the $\hat{N}_{\mathbf{i}}^{\mathbf{p}}(\mathbf{x})$:

$$N^{\mathbf{p}}(\mathbf{x}) = \sum_{\mathbf{i}} \delta_{\mathbf{i}}^{\mathbf{p}} \hat{N}_{\mathbf{i}}^{\mathbf{p}}(\mathbf{x}), \ \mathbf{i} \in \text{refined grid.}$$

The coefficients $\delta_{\mathbf{i}}^{\mathbf{p}}$ can be calculated by

$$\delta_{\mathbf{i}}^{\mathbf{p}} = \frac{1}{p^{m-r}} \mu_{\mathbf{i}}^{\mathbf{p}} ,$$

together with the recursion

$$\mu_{\mathbf{i}}^{\mathbf{p}} = \sum_{\ell=0}^{p-1} \mu_{\mathbf{i}-\ell\mathbf{k}}^{\mathbf{p}-\mathbf{k}} , \text{ any } \mathbf{k} \lneq \mathbf{p} \text{ (component wise)},$$

where

$$\mu_{\mathbf{i}}^{\mathbf{o}} = \begin{cases} 1 \text{ if } \mathbf{i} = \mathbf{o}, \\ 0 \text{ otherwise.} \end{cases}$$

This recursion was first found by Prautzsch in 1983 for p=2 (Prautzsch [11]). We give a geometric proof that is, likewise, due to Prautzsch.

Let μ_i^p be the number of subboxes that generate the same subbox spline $\hat{N}_i^p(x)$. Then from the normalization of $\hat{N}_i^p(x)$ one gets

$$\delta_i^p = \frac{1}{p^{m-r}} \mu_i^p .$$

Now we consider the p layers of subboxes in any direction, as shown in Figure 14.

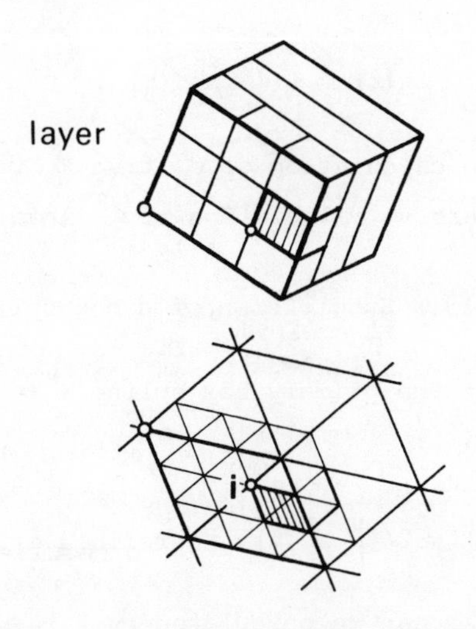

Fig. 14. The layer by Prautzsch

It is clear that the number of first layer subboxes which generate the same $\hat{N}_i^p(x)$ equals the number μ_i^{p-k} of box faces which generate the same lower degree box spline $\hat{N}_i^{p-k}$, where k corresponds to the direction of the layers. Summing up the layers gives the recursion formula above.

This recursion formula was independently developed also by Cohen, Lyche and Riesenfeld and by Dahmen and Micchelli in 1984 from an integral formula (Dahmen et al. [9], Cohen et al. [7]).

It is clear, how to extend the composition to translates of box splines. By the use of $N_j^p(x) = N^p(x-j)$ one gets

$$N_j^p(x) = \sum_i \delta_{i-pj}^p \hat{N}_i^p(x) .$$

9. REFINEMENT OF THE CONTROL NET

Let j be the points of the initial grid and let i be the points of the refined grid. As in Section 6 we consider linear combinations of translates of a box spline

$$s(x) = \sum_j d_j^p N_j^p(x) ,$$

which can also be written

$$s(x) = \sum_i \hat{d}_i^p \hat{N}_i^p(x) ,$$

where the recursive calculation of N_j^p from $\hat{N}_i^p$ implies the following dual recursive calculation of $\hat{d}_i^p$ from d_j^p:

$$\hat{d}_i^q = \sum_{\ell=0}^{p-1} \hat{d}_{i-\ell k}^{q-k}$$

$$\hat{d}_i^o = \begin{cases} \dfrac{1}{p^{m-r}} d_i^p & \text{if } i,j \text{ are coincident} \\ o & \text{otherwise.} \end{cases}$$

This control net refinement was first given for some special cases of m by Boehm in 1983, for p=2 by Prautzsch in 1983 and more generally by Dahmen and Micchelli in 1984 (Boehm [2], Prautzsch [11], Dahmen et al. [8]).

As an example we consider translates of a box spline with $\mathbf{p}$ = [1,2,1] (see Fig. 2) and a 3-partition, i.e. p = 3. Fig. 15 shows the schemes of control points, i.e. the rows of their coordinates. The initial scheme is the scheme of the given d_j^p, its refined scheme will be filled with zeros. The following schemes are smoothed by using the key

which means, one has to write the sum of p = 3 points at the position of the last (into a new scheme). This should be done in the m directions given by $\mathbf{p}$, $\mathbf{p}$ = [1,2,1] here. The elements of the final scheme are to be divided by p^{m-r}.

A numerical example is shown in Fig. 18.

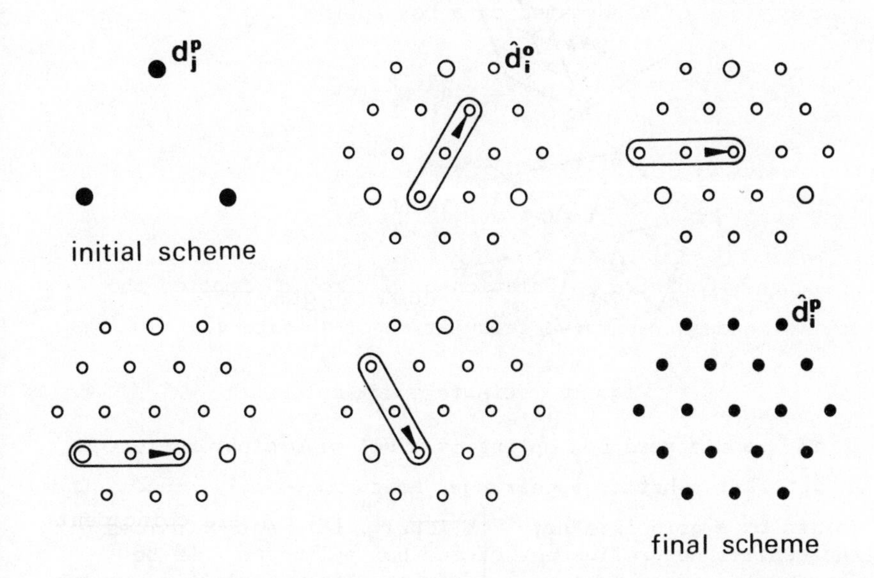

Fig. 15. Refinement of the control net, schemes

10. DISCRETE BOX SPLINES

One can condense the mass of a box to its centre of gravity. This centre is invariant under affine mappings.

Let $\hat{c}_i^p$ be the centre of the support of $\hat{N}_i^p$ forming a new refined grid corresponding to the grid $\mathbf{i}$. For large p, it is obvious that $N^p_{(x)}$ is approximated at $x = \hat{c}_i^p$ by the normalized number δ_i^p of subboxes whose centre falls into $\hat{c}_i^p$. This approximation is called a "discrete box spline", and it depends on p.

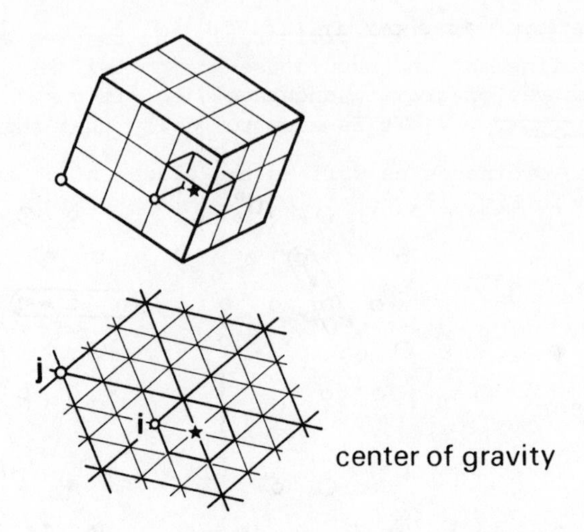

Fig. 16. Discrete box splines

Thus a discrete box spline is a set of ordinates erected at $\hat{c}_i^p$. For a better appearance, however, we will connect the points by a grid like net. Obviously, this net coincides with the control net refinement of one box spline and can be constructed as such a net. In particular, the initial control net, where p = 1, forms a pyramid. Figure 17 shows two discrete box spline nets belonging to the quadratic box spline of Fig. 2 for p = 1 and p = 3.

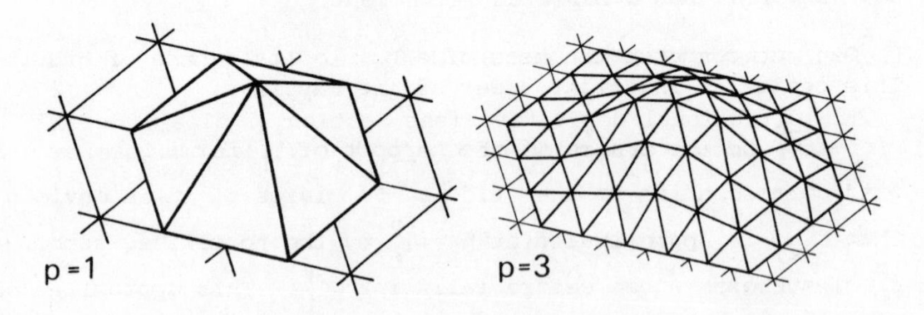

Fig. 17. Discrete box spline nets

Certainly, the refinement algorithms of Section 9 will construct the discrete box spline net (as a generalization of

Greville's theorem) in all coordinates. But to give a clear idea of the refinement one should use a symmetric key that leaves the centres of gravity unchanged, eg. if $p = 3$ use the key . It is easy to verify that this key constructs the ordinates as well as the placement of the net points, shown in Fig. 17.

```
  0   0                  0 0 0 0 0 0
                        0 1 2 3 2 1 0
                       0 1 3 5 5 3 1 0
0   1   0            ( 0 1 3 6 7 6 3 1 0 )/9
                       0 1 3 5 5 3 1 0
                        0 1 2 3 2 1 0
  0   0   p=1            0 0 0 0 0 0        p=3
```

Fig. 18. Discrete box splines

Dahmen and Micchelli, as well as Cohen, Lyche and Riesenfeld erect the ordinates $\delta_{\mathbf{i}}^{\mathbf{p}}$ at the points $\mathbf{i}$, therefore constructing a somewhat translated discrete box spline. Clearly, this unsymmetric form converges for repeated refinements to the box spline also (Dahmen et al. [9], Cohen et al. [7]).

As a generalization of a discrete box spline, the refined net of control points of a spline can be viewed as a "discrete spline". It is obvious that this discrete spline will converge to the spline for a repeated refinement.

11. DIFFERENTIATION OF BOX SPLINES

The directional derivative (see Section 3) of a function $\mathbf{s}(\mathbf{x})$ at $\mathbf{x}$ in the direction of $\mathbf{v} = \alpha_1 \mathbf{u}_1 + \ldots + \alpha_n \mathbf{u}_n$ can be written

$$D_{\mathbf{v}} \mathbf{s}(\mathbf{x}) = (\alpha_1 D_1 + \ldots + \alpha_n D_n)\, \mathbf{s}(\mathbf{x}),$$

where D_1 is the directional derivative in direction of $\mathbf{u}_1$ etc. Furthermore, if $\mathbf{s}(\mathbf{x})$ is a spline then

$$D_k \mathbf{s}(\mathbf{x}) = \sum_{\mathbf{i}} \mathbf{d}_{\mathbf{i}} D_k N_{\mathbf{i}}^{\mathbf{p}}(\mathbf{x}).$$

Therefore it is sufficient to consider the directional derivative of a box spline in only one direction of the grid:

$$D_k N_i^p(x) = N_i^{p-k}(x) - N_{i-k}^{p-k}(x).$$

Hence the directional derivative of the spline is a spline of order $p-k$

$$D_k s(x) = \sum_i d_i^k N_i^{p-k}(x)$$

The derivatives of $N_i^p(x)$ correspond to the dual construction of the control net

$$d_i^k = d_{i+k} - d_i ,$$

as illustrated in Figure 19.

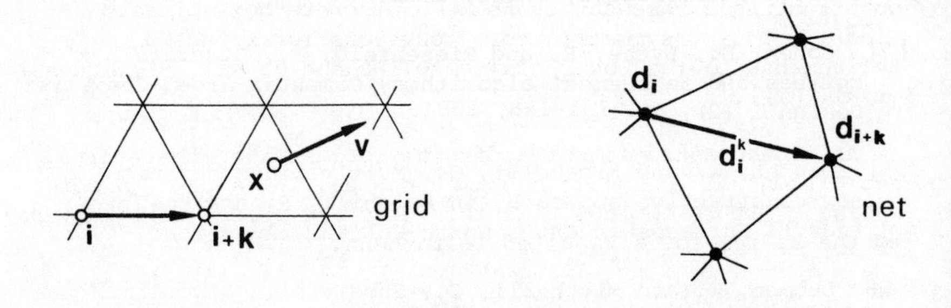

Fig. 19. Differentation of box splines

12. TENSOR PRODUCT SPLINES

If the grid i under consideration is the direct sum of two grids $i = i_1 + i_2$, such that $x = x_1 + x_2$ and $p = p_1 + p_2$, but span $i_1 \cap$ span $i_2 = 0$, it is easy to verify that

$$N_i^p(x) = N_{i_1}^{p_1}(x_1)\, N_{i_2}^{p_2}(x_2).$$

The corresponding linear combinations are called "tensor product splines".

13. REFERENCES

[1] Boehm, W. , Inserting new knots into B-spline curves, Computer Aided Design 12 (4), pp. 199-201, 1980.

[2] Boehm, W., Subdividing multivariate splines, Computer Aided Design 15, pp. 345-352, 1983

[3] Boehm, W., Farin, G. and Kahmann, J., A survey of curve and surface methods in CAGD, Computer Aided Geometric Design 1, pp. 1-60, 1984.

[4] Boehm, W., Calculating with boxsplines, Computer Aided Design 1 (2), pp. 149-162, 1984.

[5] de Boor, C. and Hoellig, K., B-splines from parallelepipeds, Journal d'Analyse Math., 42, pp. 99-115, 1982/83.

[6] de Casteljau, P., Courbes et surfaces a pôles, André Citroën automobilies, Paris, 1963.

[7] Cohen, E., Lyche, R. and Riesenfeld, R., Discrete box splines and refinement algorithms, Computer Aided Geometric Design 1 (2), pp. 131-148, 1984.

[8] Dahmen, W, and Micchelli, C., Multivariate splines - a new constructive approach, in Barnhill, R. and Boehm, W. (eds.): *Surfaces in CAGD,* North-Holland, Amsterdam, 1983.

[9] Dahmen, W. and Micchelli, C., Subdivision algorithms for the generation of box spline surfaces, Computer Aided Geometric Design 1 (2), pp. 115-129, 1984.

[10] Farin, G.E., Subsplines über dreiecken, Dissertation Technische Universitaet Braunschweig, 1979.

[11] Prautzsch, H., Unterteilungsalgorithmen für multivariate Splines - en geometrischer Zugang, Dissertation Technische Universitaet Braunschweig, 1983/84.

[12] Sabin, M.A., The use of piecewise forms for the numerical representation of shape, Dissertation Hungarian Academy of Sciences, Budapest, 1977.

N-SIDED SURFACE PATCHES

J.A. Gregory
(Brunel University)

1. INTRODUCTION

A surface in computational geometry is frequently represented in a piecewise defined, vector valued, parametric form, where the pieces, or 'patches' of the surface have rectangular domains of definition. More general polygonal patch representations are needed, however, which can be used within a rectangular patch framework. For example, Figure 1 shows the boundary curves, or 'wire frame', of a model problem which requires a triangular patch representation.

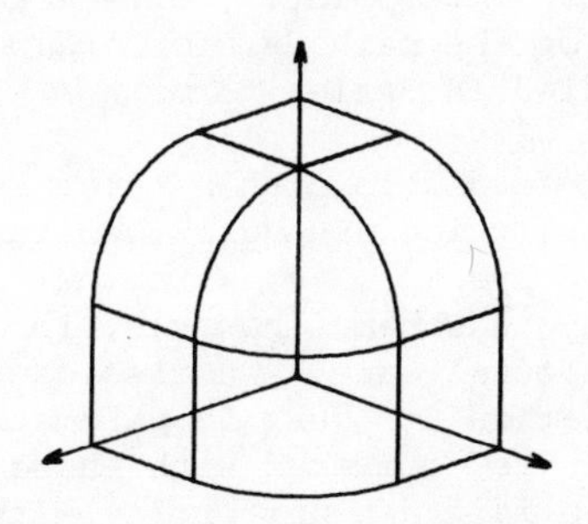

Fig. 1. A model problem

There is also a new interest in the complete representation of vector valued parametric surfaces by piecewise defined triangular patches. Although there is a large body of knowledge concerning such representations for scalar valued functions, it is not obvious that this theory can be easily adapted to treat the vector valued case. The problem common to both triangular and, more generally, polygonal vector valued surface patches, is that the domains of the patches surrounding a given patch cannot be assumed to belong to a common planar domain. Thus continuity considerations between patches must be treated in vector terms, rather than relying on scalar continuity assumptions between components of the vector form.

In this paper 'blending function' interpolation methods are considered for the representation of vector valued surfaces defined over polygonal domains. The term 'blending function' means that the surface is constructed as a blend of simpler constituents. The interpolation methods are also 'transfinite', a term used by W. Gordon to denote interpolation to an infinite set of data.

For the purposes of mathematical description let

$$\underline{f}(V) = [f_1(V), f_2(V), f_3(V)], \quad V \in \Omega, \tag{1.1}$$

denote a surface in three dimensional Euclidean space, defined on a polygonal domain Ω. Then a blending function interpolant, denoted by

$$\underline{p}(V) = [p_1(V), p_2(V), p_3(V)], \quad V \in \Omega, \tag{1.2}$$

is to be constructed, such that it matches $\underline{f}$ and its tangent plane at all points on the boundary of Ω. In practice the data $\underline{f}$ will not be given everywhere on Ω. Instead a boundary curve and cross-boundary tangent vector are defined along each side of the polygon. Also interpolatory data are not necessary for the implementation of the methods, for example the boundary data might be supplied in Bezier form.

It is convenient to distinguish between representations defined on triangles, which are described in Section 2 and those defined on N-sided polygons, N>4, which are described in Section 3. The polygon schemes however, include schemes for the triangle as a special case. We also use a slightly different index notation for certain parametric variables defined on the triangle, compared with those used on the polygon. This is to maintain uniformity with the existing literature on triangular patches, although here it leads to an inconsistency of notation between Sections 2 and 3.

2. TRIANGULAR SURFACE PATCHES

In this section two types of blending function interpolants are considered, which define vector valued surfaces on a triangular domain.

2.1 Notation

Let Ω be an equilateral triangle of height unity with vertices V_i, $i = 1,2,3$. The coordinates of a general point $V \in \Omega$ and the vertices V_i, $i = 1,2,3$ are not required. Instead parameters λ_i, $i = 1,2,3$, are employed, where λ_i denotes the perpendicular distance of V from the side opposite V_i, see Figure 2. We then have

$$\lambda_1 + \lambda_2 + \lambda_3 = 1 \tag{2.1}$$

and

$$V = \lambda_1 V_1 + \lambda_2 V_2 + \lambda_3 V_3. \tag{2.2}$$

(The parameters λ_i, $i = 1,2,3$, correspond to the barycentric or areal coordinates of a general triangle.)

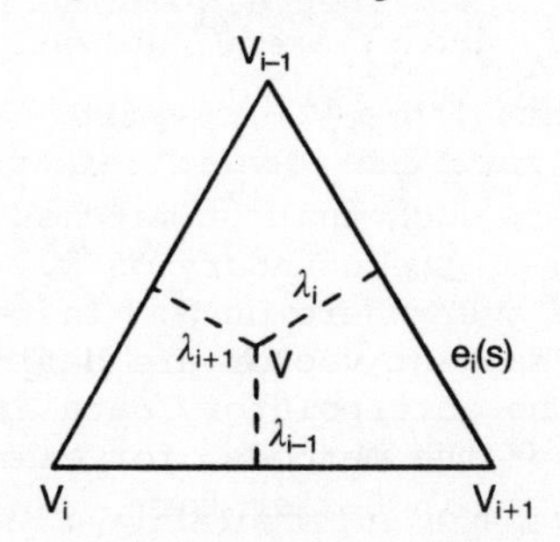

Fig. 2. The triangular domain Ω

Let the side corresponding to $\lambda_i = 0$ be defined by the equation

$$e_i(s) = (1-s)\, V_{i+1} + s\, V_{i-1}, \quad 0 \leq s \leq 1. \tag{2.3}$$

(Here an index is equivalent to the index 1, 2, or 3 if the two indices are congruent modulo 3.) A boundary curve

$$\underline{f}_i(s) = [f_{1,i}(s), f_{2,i}(s), f_{3,i}(s)] \tag{2.4}$$

and a cross-boundary tangent vector

$$\underline{t}_i(s) = [t_{1,i}(s), t_{2,i}(s), t_{3,i}(s)] \tag{2.5}$$

can then be defined with respect to s on the side e_i, see Figure 3.

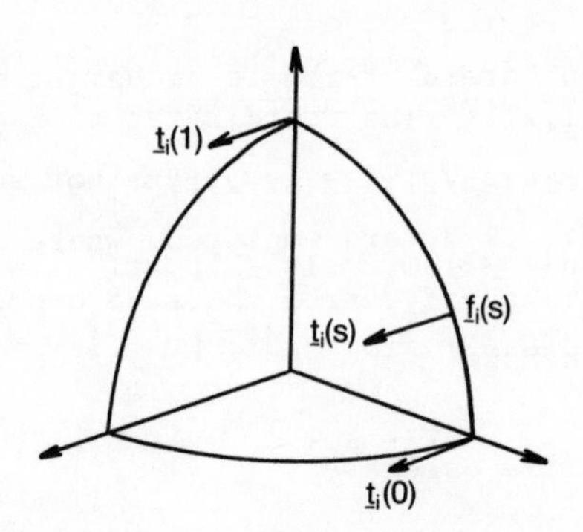

Fig. 3. The range $\underline{f}$

(Here s increases in a positive, i.e. anti-clockwise, direction on each side where an anti-clockwise ordering of the vertices is assumed.)

We assume that $\underline{f}_i$ and $\underline{t}_i$ are C^1 univariate vector functions which are consistent with a C^1 bivariate function $\underline{f}$, with domain Ω, where we make the identifications

$$\underline{f}(e_i(s)) = \underline{f}_i(s) \text{ and } \underline{f}_{n_i}(e_i(s)) = \underline{t}_i(s). \tag{2.6}$$

The notation $\underline{f}_{n_i}$ denotes a derivative of $\underline{f}$ along an inward direction $n_i(s)$ to the side $e_i(s)$. The precise definition of $n_i(s)$ will depend on the interpolation scheme and is not necessarily of practical concern. We assume, however, that

$$n_i(0) = V_i - V_{i+1} \text{ and } n_i(1) = V_i - V_{i-1}, \tag{2.7}$$

which gives the necessary conditions

$$\underline{t}_i(0) = -\dot{\underline{f}}_{i-1}(1) \text{ and } \underline{t}_i(1) = \dot{\underline{f}}_{i+1}(0) \tag{2.8}$$

on the specified data, see Figure 3.

If the triangular patch occurs within a rectangular patch framework, then the boundary data must be consistent with that of the adjoining rectangular patches. Conditions (2.8) then arise in a natural way from the equivalent conditions on the rectangular patches.

The above discussion is illustrated by the following simple model problem

A Model Problem Let the surface of Figure 3 be the octant of a sphere of unit radius, centre the origin. The appropriate boundary curves and cross-boundary tangencies are given by

$$\underline{f}_1(s) = [0,\cos(s\pi/2),\sin(s\pi/2)], \quad \underline{t}_1(s) = [\pi/2,0,0]$$

$$\underline{f}_2(s) = [\sin(s\pi/2),0,\cos(s\pi/2)], \quad \underline{t}_2(s) = [0,\pi/2,0]$$

$$\underline{f}_3(s) = [\cos(s\pi/2),\sin(s\pi/2),0] \quad \underline{t}_3(s) = [0,0,\pi/2].$$

2.2 The Brown-Little Schemes

An interpolation scheme of Brown and Little is described in Barnhill [1], which matches a scalar valued function, and its first derivatives, on the sides of a triangular domain. The scheme is a convex combination of three constituent interpolants, each of which match data on just one of the sides of the triangle. The original Brown-Little scheme uses interpolants defined along directions normal to the sides of the triangular domain. Nielson [8] describes a modified Brown-Little scheme which uses interpolants defined along radial directions from a side to its opposite vertex, see also Little [7]. This 'side-vertex' approach is appropriate to the vector valued problem which we now describe here.

Let $E_i = E_i(V)$ denote the intersection of the radial line through V_i and V with the side e_i, see Figure 4. Then

$$E_i = [\lambda_{i+1}/(1-\lambda_i)]V_{i+1} + [\lambda_{i-1}/(1-\lambda_i)]V_{i-1} \qquad (2.9)$$

where $V = \lambda_{i-1}V_{i-1} + \lambda_i V_i + \lambda_{i+1}V_{i+1}$, see (2.2).

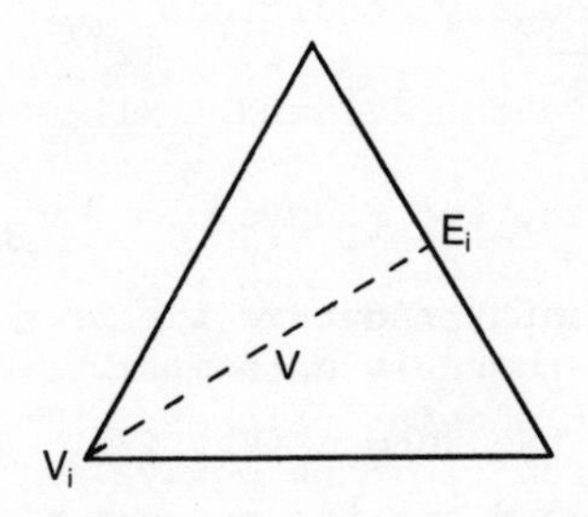

Fig. 4. The radial construct

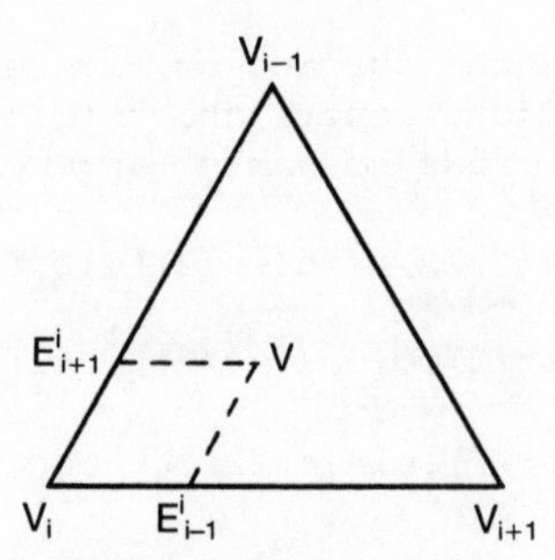

Fig. 5. The parallel to sides construct

Since

$$V = (1-\lambda_i)E_i + \lambda_i V_i, \quad 0 \le \lambda_i \le 1,$$

then bivariate interpolants can be constructed, by use of univariate interpolation in the variable λ_i along the radial direction $n_i = V_i - E_i$. In particular, functions $\underline{p}_i(V)$, which match $\underline{f}$ and its tangent plane on the side e_i, are given by the following choices of linear Taylor, quadratic Hermite, and cubic Hermite interpolation respectively:

$$\underline{p}_i(V) = \underline{f}(E_i) + \lambda_i \underline{f}_{n_i}(E_i), \tag{2.10}$$

$$\underline{p}_i(V) = (1-\lambda_i^2)\underline{f}(E_i) + \lambda_i(1-\lambda_i)\underline{f}_{n_i}(E_i) + \lambda_i^2\underline{f}(V_i), \tag{2.11}$$

$$\underline{p}_i(V) = (1-\lambda_i)^2(2\lambda_i+1)\underline{f}(E_i) + (1-\lambda_i)^2\lambda_i\underline{f}_{n_i}(E_i)$$
$$+ \lambda_i^2(-2\lambda_i+3)\underline{f}(V_i) + \lambda_i^2(\lambda_i-1)\underline{f}_{n_i}(V_i). \tag{2.12}$$

The final interpolant is defined by the convex combination

$$\underline{p}(V) = \sum_{i=1}^{3} \alpha_i(V)\, \underline{p}_i(V),$$

where

$$\alpha_i(V) = \lambda_{i-1}^2 \lambda_{i+1}^2 / (\lambda_2^2\lambda_3^2 + \lambda_1^2\lambda_3^2 + \lambda_1^2\lambda_2^2). \tag{2.14}$$

The weights sum to unity and have the property that on the sides e_{i-1} and e_{i+1} there is no contribution to $\underline{p}$, or its tangent plane, from the term $\alpha_i(V)\underline{p}_i(V)$. These properties ensure that $\underline{p}$ matches $\underline{f}$ and its tangent plane on all sides of the triangle. This argument has ignored the singularities at the vertices V_i but these can be shown to be removable in the sense that $\underline{p}$ matches $\underline{f}$, and its tangent plane in the limit $V \to V_i$, $V \in \Omega$.

Remark The radial cubic Hermite (2.12) will match all boundary curves in the case where these curves also have cubic form. The quadratic terms in the weights (2.14) can then be replaced by linear forms, e.g. see Lawson [6].

To implement the scheme, using either of (2.10), (2.11) or (2.12), the correct identifications with the given boundary data must be made. Since $E_i = e_i(s_i)$, where

$$s_i = \lambda_{i-1}/(1-\lambda_i),$$

we have

$$\underline{f}(E_i) = \underline{f}_i(s_i) \text{ and } \underline{f}_{n_i}(E_i) = \underline{t}_i(s_i).$$

Also

$$\underline{f}(V_i) = \underline{f}_{i-1}(0) = \underline{f}_{i+1}(0).$$

Finally the definition $n_i = V_i - E_i$ leads to the form

$$\underline{f}_{n_i}(V_i) = -[\lambda_{i+1}/(1-\lambda_i)]\underline{t}_{i+1}(1) - [\lambda_{i-1}/(1-\lambda_i)]\underline{t}_{i-1}(0).$$

The theory of the Brown-Little schemes, with the linear interpolants (2.10) is illustrated in Figures 6, 7 and 8, where the boundary data for the triangular patch correspond to that for the octant of a sphere.

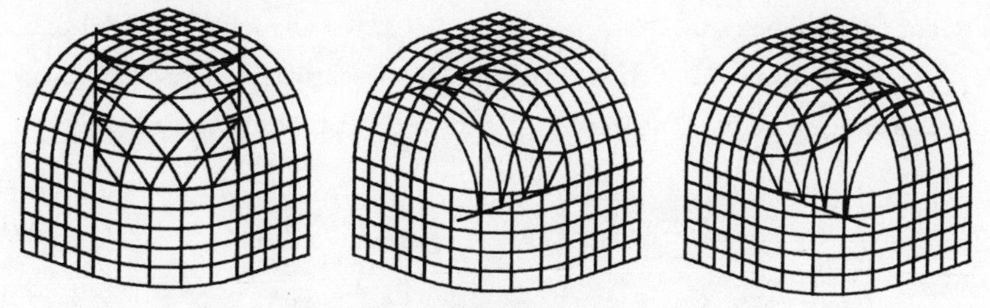

Fig. 6. The linear interpolants $\underline{p}_i$, i=1,2,3

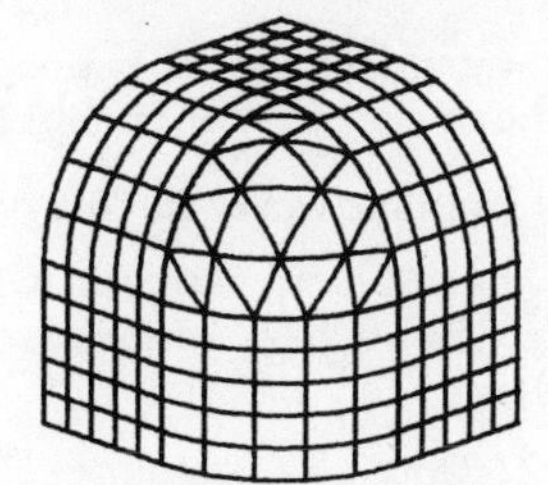

Fig. 7. The convex combination

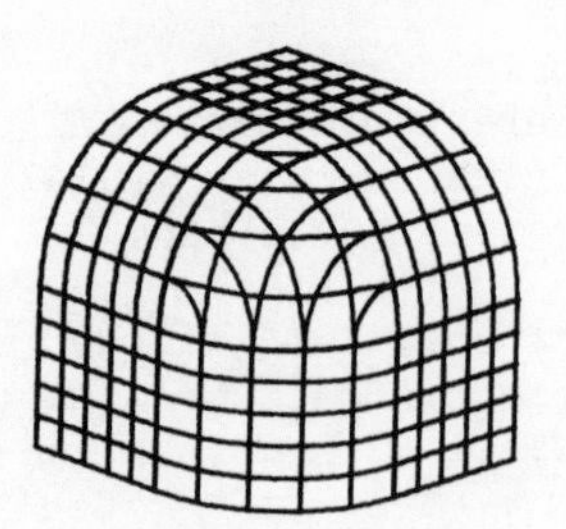

Fig. 8. Cross-sections

2.3 The Gregory-Charrot Scheme

The Gregory-Charrot interpolant [5], is a vector valued adaptation of a scalar valued scheme given in Gregory [4]. The scheme is a convex combination of three constituent interpolants each of which match data on two sides of the triangle. The two-sided interpolants are constructed using the Boolean sum ideas formalised by Gordon [3].

Let

$$E^i_{i+1} = (1-\lambda_{i-1})V_i + \lambda_{i-1}V_{i-1} \quad (2.15)$$

be the intersection of the side e_{i+1} with the line through V parallel to e_{i-1}. Similarly let

$$E^i_{i-1} = (1-\lambda_{i+1})V_i + \lambda_{i-1}V_{i+1} \quad (2.16)$$

be the intersection of e_{i-1} with the line through V parallel to e_{i+1}, see Figure 5. Then linear Taylor interpolants can be defined, in the variables λ_{i+1} and λ_{i-1} respectively, by

$$\underline{T}^i_{i+1}(V) = \underline{f}(E^i_{i+1}) + \lambda_{i+1}\underline{f}_{n_{i+1}}(E^i_{i+1}) , \quad (2.17)$$

$$\underline{T}^i_{i-1}(V) = \underline{f}(E^i_{i-1}) + \lambda_{i-1}\underline{f}_{n_{i-1}}(E^i_{i-1}), \quad (2.18)$$

where here $n_{i+1} = V_{i+1} - V_i$ and $n_{i-1} = V_{i-1} - V_i$.

A "Tensor product' interpolant can be defined from (2.17) and (2.18) by

$$\underline{T}^{i}_{i+1,i-1}(V) = [1 \quad \lambda_{i+1}] \begin{bmatrix} \underline{f}(V_i) & \underline{f}_{n_{i-1}}(V_i) \\ \underline{f}_{n_{i+1}}(V_i) & \underline{f}_{n_{i+1}n_{i-1}}(V_i) \end{bmatrix} \begin{bmatrix} 1 \\ \lambda_{i-1} \end{bmatrix}, \tag{2.19}$$

where $\underline{f}_{n_{i+1},n_{i-1}} = \partial^2 \underline{f}/\partial n_{i+1}\,\partial n_{i-1}$. The dual Tensor product form $\underline{T}^{i}_{i-1,i+1}(V)$ involves $\underline{f}_{n_{i-1},n_{i+1}}(V_i)$ and is identical with $T^{i}_{i+1,i-1}(V)$ if

$$\underline{f}_{n_{i+1},n_{i-1}}(V_i) = \underline{f}_{n_{i-1},n_{i+1}}(V_i). \tag{2.20}$$

This is called the 'twist' compatibility condition and data which satisfy this condition at each vertex are said to be compatible.

The 'Boolean sum' interpolant is defined by

$$\underline{p}_i(V) = \underline{T}^{i}_{i+1}(V) + \underline{T}^{i}_{i-1}(V) - \underline{T}^{i}_{i+1,i-1}(V) \tag{2.21}$$

and has the property that, for compatible data, it matches $\underline{f}$, and its tangent plane, on e_{i+1} and e_{i-1}. If the data are not compatible, the rational generalization

$$\underline{p}_i(V) = \underline{T}^{i}_{i+1}(V) + \underline{T}^{i}_{i-1}(V) - [\lambda_{i-1}/(\lambda_{i+1} + \lambda_{i-1})]\underline{T}^{i}_{i+1,i-1}(V)$$
$$- [\lambda_{i+1}/(\lambda_{i+1} + \lambda_{i-1})]\underline{T}^{i}_{i-1,i+1}(V) \tag{2.22}$$

can be used. The Boolean sum interpolant also has the property that it has cubic precision, since it has the union of the precisions of $\underline{T}^{i}_{i+1}(V)$ and $\underline{T}^{i}_{i-1}(V)$.

The final interpolant is defined by

$$\underline{p}(V) = \sum_{i=1}^{3} \alpha_i(V)\, \underline{p}_i(V), \tag{2.23}$$

where the appropriate weights are given by

$$\alpha_i(V) = \lambda_i^2/(\lambda_1^2 + \lambda_2^2 + \lambda_3^2) \quad \textit{(rational)} \tag{2.24}$$

or

$$\alpha_i(V) = \lambda_i^2(3 - 2\lambda_i + 6\lambda_{i+1}\lambda_{i-1}) \quad \textit{(polynomial)} \qquad (2.25)$$

The weights sum to unity and the factor λ_i^2 implies that $\alpha_i(V)\underline{p}_i(V)$ does not contribute to $\underline{p}$ or its tangent plane on e_i. These properties ensure that $\underline{p}$ matches $\underline{f}$, and its tangent plane, on the entire boundary of the triangle. The cross boundary direction for the final scheme is now

$$n_i(s) = \alpha_{i-1}(e_i(s))(V_i - V_{i-1}) + \alpha_{i+1}(e_i(s))(V_i - V_{i+1}).$$

Since $E^i_{i+1} = e_{i+1}(1-\lambda_{i-1})$ and $E^i_{i-1} = e_{i-1}(\lambda_{i+1})$, the scheme is implemented with

$$\underline{f}(E^i_{i+1}) = \underline{f}_{i+1}(1-\lambda_{i-1}),\ \underline{f}_{n_{i+1}}(E^i_{i+1}) = \underline{t}_{i+1}(1-\lambda_{i-1}),$$

$$\underline{f}(E^i_{i-1}) = \underline{f}_{i-1}(\lambda_{i+1}), \quad \underline{f}_{n_{i-1}}(E^i_{i-1}) = \underline{t}_{i-1}(\lambda_{i+1}).$$

The twist terms are given by

$$\underline{f}_{n_{i+1},n_{i-1}}(V_i) = \dot{\underline{t}}_{i-1}(0),\ \underline{f}_{n_{i-1},n_{i+1}}(V_i) = \dot{\underline{t}}_{i+1}(1),$$

these being equal for compatible data.

The theory of the Gregory-Charrot scheme is illustrated in Figures 9, 10 and 11.

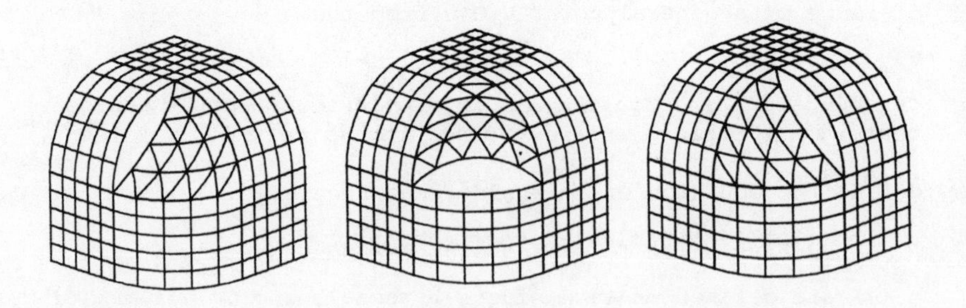

Fig. 9. The Boolean sum interpolants $\underline{p}_i$, i=1,2,3

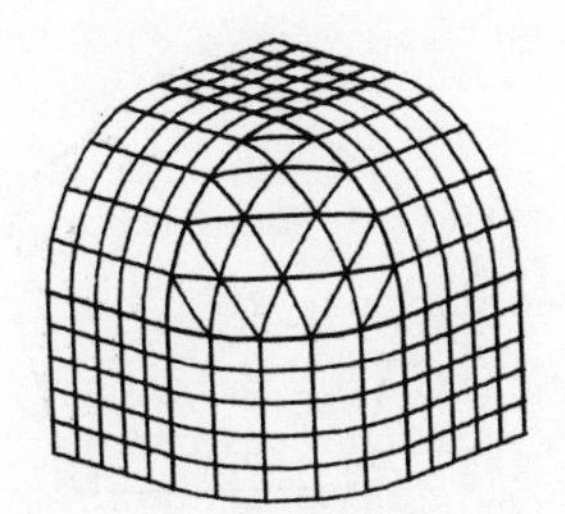

Fig. 10. The convex combination

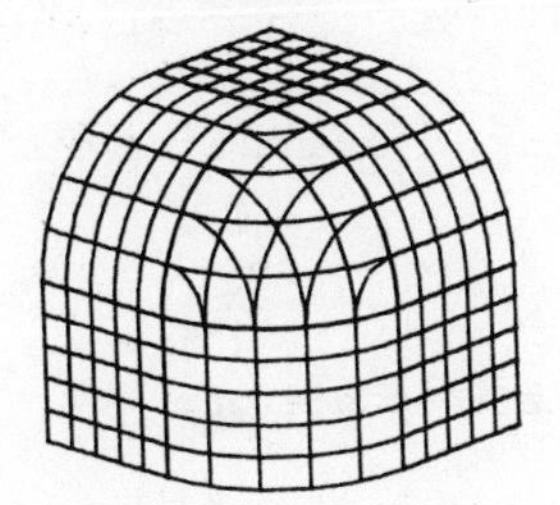

Fig. 11. Cross-sections

3. POLYGONAL SURFACE PATCHES

The basic ideas of the previous section are now extended to treat the polygonal patch problem.

3.1 Notation

Let Ω be a regular polygon with N sides of length d and vertices V_i, $i = 0,\ldots,N-1$. Let λ_i denote the perpendicular distance of a general point $V \in \Omega$ from the side joining V_i and V_{i+1}, and let

$$e_i(s) = (1-s)V_i + s\,V_{i+1},\ 0 \leq s \leq 1, \tag{3.1}$$

define the equation of this side, see Figure 12.

A boundary curve $\underline{f}_i(s)$ and a cross-boundary tangent vector $\underline{t}_i(s)$ are defined on each side e_i, these being C^1 univariate functions which are assumed to be consistent with a C^1 bivariate function $\underline{f}$, where

$$\underline{f}(e_i(s)) = \underline{f}_i(s) \text{ and } \underline{f}_{n_i}(e_i(s)) = \underline{t}_i(s).$$

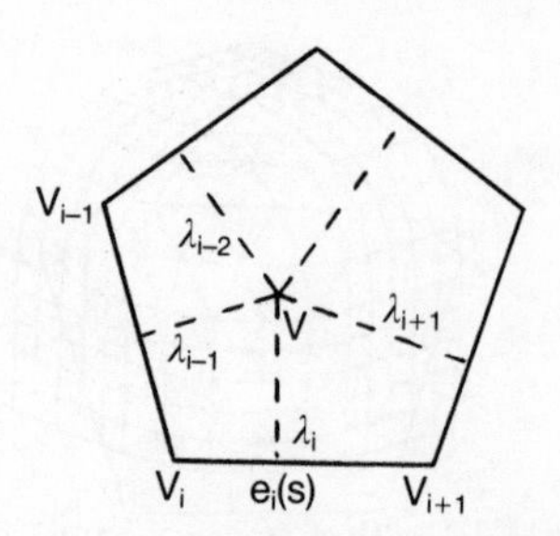

Fig. 12. The polygonal domain Ω

Here $n_i(s)$ is an inward direction to the side $e_i(s)$ such that

$$n_i(0) = V_{i-1} - V_i, \; n_i(1) = V_{i+2} - V_{i+1}. \tag{3.2}$$

This gives the necessary conditions

$$\underline{t}_i(0) = -\dot{\underline{f}}_{i-1}(1), \; \underline{t}_i(1) = \dot{\underline{f}}_{i+1}(0) \tag{3.3}$$

on the specified data.

Two blending function schemes for interpolating the data are now considered.

3.2 The Brown-Little Scheme

Consider the radial line joining V with the point of intersection of the sides e_{i-1} and e_{i+1} and let the intersection of this line with the side e_i be E_i, see Figure 13. Then

$$E_i = \frac{\lambda_{i+1}}{\lambda_{i+1} + \lambda_{i-1}} V_i + \frac{\lambda_{i-1}}{\lambda_{i+1} + \lambda_{i-1}} V_{i+1}. \tag{3.4}$$

Let

$$d = 1/\sin\theta, \; \theta = 2\pi/N \tag{3.5}$$

define the length of a side of the polygon. Then λ_i is an appropriate variable for linear Taylor interpolation along the radial direction given by

$$\underline{p}_i(V) = \underline{f}(E_i) + \lambda_i \underline{f}_{n_i}(E_i). \tag{3.6}$$

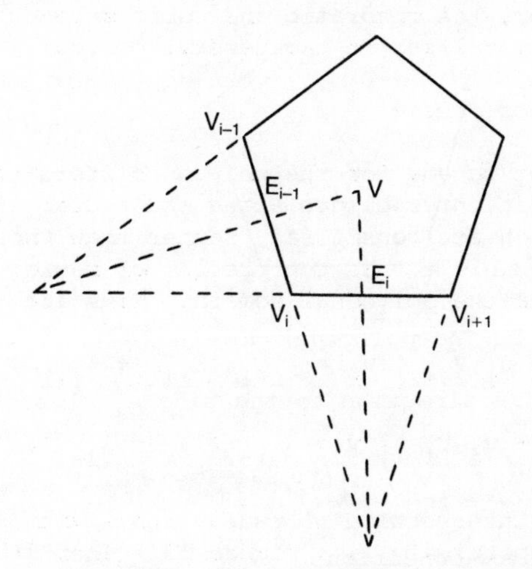

Fig. 13. The radial construct

Here

$$n_i = [\lambda_{i+1}/(\lambda_{i+1}+\lambda_{i-1})](V_{i-1}-V_i) + [\lambda_{i-1}/(\lambda_{i+1}+\lambda_{i-1})](V_{i+2}-V_{i+1})$$

denotes the radial direction.

The final interpolant is defined by the convex combination

$$\underline{p}(V) = \sum_{i=0}^{N-1} \alpha_i(V)\, \underline{p}_i(V), \tag{3.7}$$

where

$$\alpha_i(V) = \prod_{j\neq i} \lambda_j^2 \Big/ \sum_{k=0}^{N-1} \prod_{j\neq k} \lambda_j^2. \tag{3.8}$$

The scheme is implemented in terms of the given boundary data with the identifications

$$\underline{f}(E_i) = \underline{f}_i(s_i),\ \underline{f}_{n_i}(E_i) = \underline{t}_i(s_i),$$

where

$$s_i = \lambda_{i-1}/(\lambda_{i+1} + \lambda_{i-1}),$$

c.f. (3.1) and (3.4).

The scheme gives the linear Taylor method for the triangle in the case N=3. However, the quadratic and cubic schemes for the triangle cannot be generalized to the general polygon.

3.3 The Gregory-Charrot Scheme

The Gregory-Charrot scheme for the polygon differs from that on the triangle in that interpolants along the radial lines of the previous subsection are considered, rather than those along parallels to a side. This avoids the problem of having to prescribe data outside the polygonal domain. Thus let

$$\begin{aligned} E_i &= (1-u_i)V_i + u_i V_{i+1}, \quad u_i = \lambda_{i-1}/(\lambda_{i+1} + \lambda_{i-1}), \\ E_{i-1} &= (1-v_i)V_i + v_i V_{i-1}, \quad v_i = \lambda_{i-2}/(\lambda_i + \lambda_{i-2}), \end{aligned} \tag{3.9}$$

define the points of intersection of radial lines with the sides e_i and e_{i-1} respectively, see Figure 13. Then linear Taylor interpolants are defined in the variables u_i and v_i by

$$\underline{T}^i_{i-1}(V) = \underline{f}(E_{i-1}) + u_i \underline{f}_{n_{i-1}}(E_{i-1}), \tag{3.10}$$

$$\underline{T}^i_i(V) = \underline{f}(E_i) + v_i \underline{f}_{n_i}(E_i). \tag{3.11}$$

These interpolants are constructed along the directions v_i = constant and u_i = constant respectively. They correspond to interpolants along the radial directions but with different parameterization to that used in (3.6). Also, since the parameterization is independent of d, for implementation it suffices to take d = 1.

The Tensor product interpolant is

$$\underline{T}^i_{i-1,i}(V) = [1 \quad u_i] \begin{bmatrix} \underline{f}(V_i) & \underline{f}_{n_i}(V_i) \\ \underline{f}_{n_{i-1}}(V_i) & \underline{f}_{n_{i-1},n_i}(V_i) \end{bmatrix} \begin{bmatrix} 1 \\ v_i \end{bmatrix} \tag{3.12}$$

with a dual expression for $\underline{T}^i_{i,i-1}(V)$. The Boolean sum interpolant is

$$\underline{p}_i(V) = \underline{T}^i_{i-1}(V) + \underline{T}^i_i(V) - \underline{T}^i_{i-1,i}(V) \tag{3.13}$$

or the rational form

$$\underline{p}_i(V) = \underline{T}^i_{i-1}(V) + \underline{T}^i_i(V) - [v_i/(u_i+v_i)]\underline{T}^i_{i-1,i}(V)$$

$$- [u_i/(u_i + v_i)]\ \underline{T}^i_{i,i-1}(V) \qquad (3.14)$$

is used if the data are not compatible, i.e.

$$\underline{f}_{n_{i-1},n_i}(V_i) \neq \underline{f}_{n_i,n_{i-1}}(V_i).$$

The full scheme for the polygon is defined by the convex combination

$$\underline{p}(V) = \sum_{i=0}^{N-1} \alpha_i(V)\ \underline{p}_i(V) \qquad (3.15)$$

where

$$\alpha_i(V) = \prod_{j\neq i-1,i} \lambda_j^2 \Big/ \sum_{k=0}^{N-1} \prod_{j\neq k-1,k} \lambda_j^2. \qquad (3.16)$$

The scheme is implemented with

$$\underline{f}(E_{i-1}) = \underline{f}_{i-1}(1-v_i),\ \underline{f}_{n_{i-1}}(E_{i-1}) = \underline{t}_{i-1}(1-v_i),$$

$$\underline{f}(E_i) = \underline{f}_i(u_i),\quad \underline{f}_{n_i}(E_i) = \underline{t}_i(u_i),$$

etc. where the twist terms are

$$\underline{f}_{n_{i-1},n_i}(V_i) = \dot{\underline{t}}_i(0),\ \underline{f}_{n_i,n_{i-1}} = \dot{\underline{t}}_i(1).$$

Examples which illustrate this scheme for the pentagon are given in Charrot and Gregory [2]. The case N = 3 gives a triangle scheme which is different from that described in sub-section 2.3 but on the triangle the u_i and v_i have singularities (which in the final scheme will be removable).

4. SUMMARY

This paper has described two basic methods for constructing triangular and polygonal vector valued surface patches which join with position and tangent plane continuity. The first method is based on the Brown-Little construct of a convex combination of one sided interpolants. The second method is the Gregory-Charrot construct of a convex combination of two sided Boolean sum interpolants. Our experience of the Gregory-

Charrot scheme is that it gives good surfaces and for compatible data it avoids the use of singular functions. The Brown-Little scheme is easier to implement but, for the general polygon, has only linear precision. However, more accurate Brown-Little schemes can be constructed for the triangle.

ACKNOWLEDGEMENT

The author thanks T. Koukouvinos for help in the production of the computer generated plots.

REFERENCES

[1] Barnhill, R.E., Representation and approximation of surfaces. In *Mathematical Software III,* J.R. Rice (ed.), pp. 69-120, Academic Press, 1977.

[2] Charrot, P. and Gregory, J.A., A pentagonal surface patch for computer aided geometric design, *Computer Aided Geometric Design* 1, pp. 87-94, 1984.

[3] Gordon, W.J., Blending function methods of bivariate and multivariate interpolation and approximation, *SIAM J. Numer. Anal.* **8**, pp. 158-177, 1971.

[4] Gregory, J.A., A blending function interpolant for triangles. In *Multivariate Approximation,* D.C. Hanscomb (ed.), pp. 279-287, Academic Press, 1978.

[5] Gregory, J.A. and Charrot, P., A C^1 triangular interpolation patch for computer aided geometric design, *Computer Graphics and Image Processing* **13**, pp. 80-87, 1980.

[6] Lawson, C.L., C^1 surface interpolation for scattered data on a sphere, *Rocky Mountain J. Math.* **14**, pp. 177-202, 1984.

[7] Little, F.F., Convex combination surfaces. In *Surfaces in CAGD,* R.E. Barnhill and W. Boehm (eds.), pp. 99-107, North Holland, 1983.

[8] Nielson, G.M., The side vertex method for interpolation in triangles, *J. Approx. Theory* **25**, pp. 318-336, 1979.

THE SOLUTION OF A FRAME MATCHING EQUATION

A.W. Nutbourne
(Cambridge University Engineering Department)

1. INTRODUCTION

The surface torsion at a point on a surface curve is related to the torsion of the curve by the formula

$$t = \tau + \phi' \tag{1.1}$$

where t is the surface torsion

τ is the curve torsion

ϕ is the angle between the curve normal $\underline{n}$ and the surface normal $\underline{N}$

$\phi' = \dfrac{d\phi}{ds}$, where s is the arc length.

For a line of curvature $t(s) = 0$ for all s. We can therefore force any curve to be a line of curvature simply by choosing

$$\phi(s) = - \int \tau(s)\,ds + \phi(0) \tag{1.2}$$

Strangely it is only the torsion profile of the curve that is relevant - no demand is made of the curvature profile. Note that $\phi(0)$ is an arbitrary constant.

For two curves to become sides of a principal patch meeting orthogonally at corner A, see Figure 1, the vector cross product of their tangents at A defines the surface normal vector $\underline{N}$ at A and this determines the angle $\phi(0)$ for each curve. The arbitrary constants are thus determined.

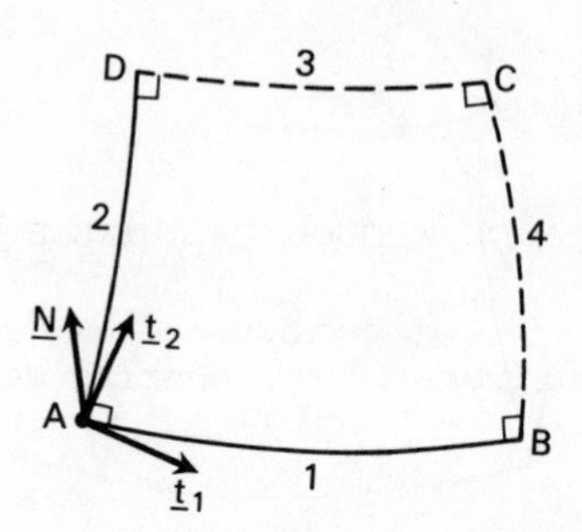

Fig. 1 Sides 1 & 2 of a Principal Patch

For the special case where a line of curvature is a plane curve, $\tau(s) = 0$ for all s so (1.2) becomes

$$\phi(s) = \phi(0) \qquad (1.3)$$

The significant point is that ϕ is constant for a plane curve and the dependence on s can be deleted from the notation.

2. THE REQUIREMENTS FOR A PRINCIPAL PATCH

2.1 The Frame Matching Equation

We can regard the vectors $[\underline{t}_1, \underline{t}_2, \underline{N}]$ as the (orthonormal) frame at corner A. Writing this in matrix form as a column vector

$$F_A = \begin{bmatrix} \underline{t}_1 \\ \underline{t}_2 \\ \underline{N} \end{bmatrix} \qquad (2.1)$$

will enable us to describe frames at the corners B, C, D (see Figure 1) using rotation matrices.

We now make the assumption that sides 1 and 2 of our embryo principal patch are <u>fully known</u>. It follows that the frames at corners B and D are fully known. Corner C is not a fixed point, but sides (3) and (4) are lines of curvature, intersecting orthogonally at C.

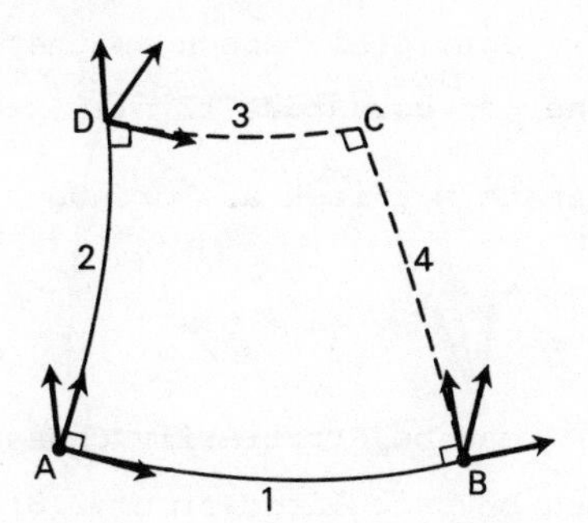

Fig. 2 Frames at Corners B, D

Suppose that a rotation matrix M is caused by <u>movement</u> along a line of curvature, then

$$F_B = M_1 F_A \tag{2.1}$$

$$F_D = RM_2 LF_A \tag{2.2}$$

where L and R are parade ground 'turn left' and 'turn right' matrices

$$L = \begin{bmatrix} 0 & 1 & 0 \\ -1 & 0 & 0 \\ 0 & 0 & 1 \end{bmatrix}, \quad R = \begin{bmatrix} 0 & -1 & 0 \\ 1 & 0 & 0 \\ 0 & 0 & 1 \end{bmatrix} \tag{2.3}$$

We require curves 3 and 4 to be lines of curvature that terminate in a common frame at corner C. Thus

$$F_C = RM_4 L\ F_B = M_3 F_D \tag{2.4}$$

Substituting for F_B and F_D from (2.1) and (2.2) gives us a frame matching equation

$$RM_4 L\ M_1 F_A = M_3 R\ M_2 L\ F_A \tag{2.5}$$

Here L and R, regarded as operators, shuffle the elements of M and change a sign here and there, but do not destroy information. It is helpful to regard RML as a "vertical" movement rotation matrix and write

$$V = RML \tag{2.6}$$

The frame matching relation is thus simplified notationally to

$$V_4 M_1 = M_3 V_2 \tag{2.7}$$

Here M_1 and V_2 are completely known as they are determined by our two given lines of curvature forming sides 1 and 2.

Since the movement matrices are orthogonal we may solve (2.7) in the form

$$V_4 = M_3 V_2 M_1^T \tag{2.8}$$

It appears that M_3 can be arbitrarily chosen and that V_4 (and hence M_4) can then be calculated from (2.8). This is fallacious, for a rather subtle reason. The M matrices are not 'any old orthogonal matrix'. They belong specifically to lines of curvature. This imposes an additional structural constraint on the elements of M. It will be shown in section 3 that M for a plane line of curvature has the structural form of a Householder orthogonal matrix, see (3.7). V_4 as calculated from (2.8) does not have the correct structural form for a line of curvature. This implies that M_3 is not arbitrary; but is constrained in some way. To examine this in detail it is better to use the balanced relation (2.7) rather than the lop-sided form (2.8). We thus regard (2.7) as the frame matching equation.

2.2 The Position Matching Equation

It is of little use getting the frames at C to match for the two paths 1 → 4 and 2 → 3 if these two paths do not meet positionally at C.

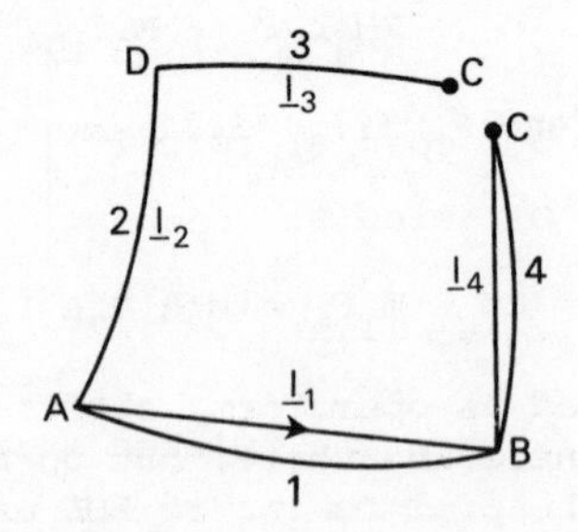

Fig. 3 Position Matching

The position matching equation is simply

$$\underline{\ell}_1 + \underline{\ell}_4 = \underline{\ell}_2 + \underline{\ell}_3 \tag{2.9}$$

where the $\underline{\ell}$ vectors are the chords for their respective sides. Sides 1 and 2 have known chord vectors so $\underline{\ell}_1$ and $\underline{\ell}_2$ are given.

Strangely, the position matching equation is quite closely linked to the frame matching equation. This is because

$$\underline{\ell} = \int \underline{t}(s)\,ds \tag{2.10}$$

for any curve. It follows that

$$\underline{\ell}_1 = \int_0^s [1\ 0\ 0]\ M_1(\sigma)F_A\ d\sigma \tag{2.11}$$

It is prudent to solve the frame matching equation first and then the position matching equation. Happily for cyclidal patches the position matching equation evaporates, as will become apparent. The frame and position matching equations are necessary constraints on the boundary curves. Martin [1] has shown that they are also sufficient to ensure that a spanning surface exists but that the lines of curvature may not all have a common functional form.

3. THE M MATRIX FOR A LINE OF CURVATURE

If a curve has an explicit rotation matrix then its M matrix follows easily if we replace $\phi(s)$ by

$$\phi(0) - \int_0^s \tau(\sigma)\,d\sigma.$$

A twisted straight line has a rotation matrix

$$\begin{bmatrix} 1 & 0 & 0 \\ 0 & \cos\beta(s) & \sin\beta(s) \\ 0 & -\sin\beta(s) & \cos\beta(s) \end{bmatrix} \tag{3.1}$$

$$\text{where } \beta(s) = \phi(s) - \phi(0) - \int \tau(s)\,ds \tag{3.2}$$

So the movement matrix when the twisted straight line is a line of curvature is

$$M(s) = \begin{bmatrix} 1 & 0 & 0 \\ 0 & 1 & 0 \\ 0 & 0 & 1 \end{bmatrix} \tag{3.3}$$

This is rather surprising result. It means that a surface frame is translated but not rotated by movement along a twisted straight line of curvature. ϕ is exactly right to compensate for the twist.

A plane curve has a rotation matrix

$$\begin{bmatrix} \cos\psi(s) & \sin\phi(0)\sin\psi(s) & \cos\phi(0)\sin\psi(s) \\ -\sin\phi(s)\sin\psi(s) & A & B \\ -\cos\phi(s)\sin\psi(s) & C & D \end{bmatrix} \tag{3.4}$$

where $A = \sin\phi(0)\sin\phi(s)\cos\psi(s) + \cos\phi(0)\cos\phi(s)$

$B = \cos\phi(0)\sin\phi(s)\cos\psi(s) - \sin\phi(0)\cos\phi(s)$

$C = \sin\phi(0)\cos\phi(s)\cos\psi(s) - \cos\phi(0)\sin\phi(s)$

$D = \cos\phi(0)\cos\phi(s)\cos\psi(s) + \sin\phi(0)\sin\phi(s)$

and $\psi(s) = \int_0^s K(\sigma)\,d\sigma$

The plane line of curvature has $\phi(s) = \phi(0)$, so its M matrix is

$$M = \begin{bmatrix} \cos\psi(s) , & \sin\phi\sin\psi(s), & \cos\phi\sin\psi(s) \\ -\sin\phi\sin\psi(s), & \cos^2\phi + \sin^2\phi\cos\psi(s), & -\sin\phi\cos\phi\,\mathrm{vers}\,\psi(s) \\ -\cos\phi\sin\psi(s), & -\sin\phi\cos\phi\,\mathrm{vers}\,\psi(s), & \sin^2\phi + \cos^2\phi\cos\psi(s) \end{bmatrix} \tag{3.5}$$

where $\mathrm{vers}\,\psi(s) = 1 - \cos\psi(s)$

If we put

$$A = [a,b,c] = [\cos\frac{\psi}{2}, \sin\phi \sin\frac{\psi}{2} , \cos\phi\sin \frac{\psi}{2}] \tag{3.6}$$

we can write this awkward trigonometric matrix as a Householder form of orthogonal matrix

$$M = G[I - 2A^TA] \tag{3.7}$$

where

$$G = \begin{bmatrix} -1 & 0 & 0 \\ 0 & 1 & 0 \\ 0 & 0 & 1 \end{bmatrix} \tag{3.8}$$

and $AA^T = a^2 + b^2 + c^2 = 1$.

We are now poised to solve the frame matching equation for plane lines of curvature.

4. SOLVING THE FRAME MATCHING EQUATION FOR PLANE LINES OF CURVATURE

Note first that M(s) for a plane line of curvature is a function of only two variables ϕ and $\psi(s)$, where ϕ is necessarily constant. Two quite different plane lines of curvature can have the same M matrix for a particular value of s provided that the area under their curvature profiles are equal for this value of s, and that they have the same ϕ. The detailed shape of their curvature profiles can be different.

We start with the Frame Matching Equation

$$RM_4LM_1F_A = M_3RM_2L\,F_A \tag{4.1}$$

Post dot with FA^T and noting that $F_A \cdot F_A^T = 1$ gives

$$RM_4LM_1 = M_3RM_2L \tag{4.2}$$

This is the frame matching equation with the frame removed, and it is possible to solve this equation with no special reference to F_A. However we prefer to introduce four unit vectors, one for each side, $\underline{e}_1$ to $\underline{e}_4$ using the initial frame in each case as a local basis.

Let

$$\underline{e}_1 = A_1F_A \tag{4.3}$$

$$\underline{e}_2 = A_2LF_A \tag{4.4}$$

$$\underline{e}_3 = A_3RM_2LF_A \tag{4.5}$$

$$\underline{e}_4 = A_4LM_1F_A \tag{4.6}$$

These vectors are each of the form AF where F is the starting frame for the side in question, see Figure 4, and A is the vector of the Householder form of the movement matrix along a side. They are all unit vectors.

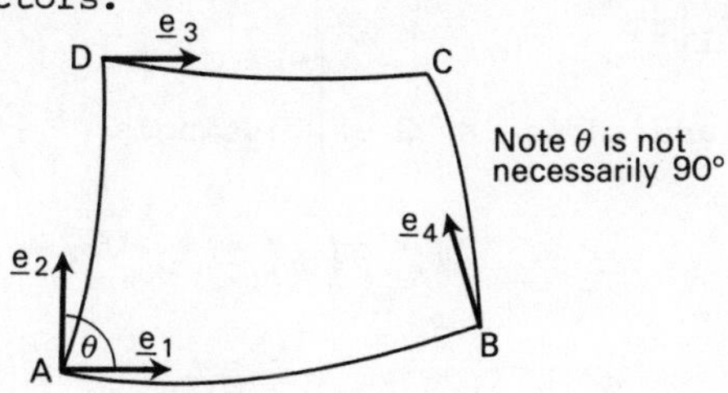

Fig. 4 Side Unit Vectors

Substituting for M_3 and M_4 in (4.1) gives

$$RG\,[I - 2A_4{}^TA_4]\,LM_1F_A = G\,[I - 2A_3^TA_3]RM_2L\,F_A$$

i.e. $$RGLM_1F_A - 2RGA_4^T\,\underline{e}_4 = GRM_2LF_{\underline{A}} - 2GA_3^T\,\underline{e}_3$$

Now substituting for M_1 and M_2 in a similar manner gives

$$RGLG\,[I - 2A_1^TA]F_A - 2RGA_4^T\,\underline{e}_4 = GRG\,[I - 2A_2^TA_2]LF_A - 2GA_3^T\,\underline{e}_3.$$

But GLG = R; GRG = L and $R^2 = L^2$ so this equation reduces to

$$R^2A_1{}^T\,\underline{e}_1 + RGA_4^T\,\underline{e}_4 = LA_2^T\,\underline{e}_2 + GA_3^T\,\underline{e}_3.$$

Multiply by L^2 and note that $L^3 = R$ to obtain

$$\begin{bmatrix} a_1 \\ b_1 \\ c_1 \end{bmatrix}\underline{e}_1 + \begin{bmatrix} b_4 \\ a_4 \\ c_4 \end{bmatrix}\underline{e}_4 = \begin{bmatrix} -b_2 \\ a_2 \\ c_2 \end{bmatrix}\underline{e}_2 + \begin{bmatrix} a_3 \\ -b_3 \\ c_3 \end{bmatrix}\underline{e}_3. \tag{4.7}$$

This is highly enigmatic, and does not yield its secrets easily! The coefficients of the vectors are so intimately connected to the vectors. The best that I can devise (and there may be some more elegant procedure) is first to show that $\underline{e}_3$ and $\underline{e}_4$ are parallel to the plane containing $\underline{e}_1$ and $\underline{e}_2$ at corner A. Call this plane the E plane and assume that $\underline{e}_3$ and $\underline{e}_4$ are not parallel, so that $\underline{e}_3 \times \underline{e}_4 \neq 0$.

Dot (4.7) with $\underline{e}_3 \times \underline{e}_4$ to obtain

$$\begin{bmatrix} a_1 \\ b_1 \\ c_1 \end{bmatrix}\underline{e}_1\cdot(\underline{e}_3 \times \underline{e}_4) = \begin{bmatrix} -b_2 \\ a_2 \\ c_2 \end{bmatrix}\underline{e}_2\cdot(\underline{e}_3 \times \underline{e}_4) \tag{4.8}$$

Squaring and adding each row, using $a^2 + b^2 + c^2 = 1$ gives

$$\underline{e}_1\cdot(\underline{e}_3 \times \underline{e}_4) = \pm\,\underline{e}_2\cdot(\underline{e}_3 \times \underline{e}_4) \tag{4.9}$$

But back substitution in (4.8) then gives

$$\begin{bmatrix} a_1 \\ b_1 \\ c_1 \end{bmatrix} = \pm \begin{bmatrix} -b_2 \\ a_2 \\ c_2 \end{bmatrix} \tag{4.10}$$

Since our given curves for sides 1 and 2 of the patch are independent, (4.10) is false. It follows that $\underline{e}_1 . (\underline{e}_3 \times \underline{e}_4)$ and $\underline{e}_2 . (\underline{e}_3 \times \underline{e}_4)$ are both zero.

If $\underline{e}_3$ is not parallel to the E plane we may use $\underline{e}_1$, $\underline{e}_2$, $\underline{e}_3$ as a basis and write

$$\underline{e}_4 = \lambda \, \underline{e}_1 + \mu \, \underline{e}_2 + \nu \, \underline{e}_3 \tag{4.11}$$

Then

$$\underline{e}_3 \times \underline{e}_4 = \lambda \, \underline{e}_3 \times \underline{e}_1 + \mu \, \underline{e}_3 \times \underline{e}_2 \tag{4.12}$$

Therefore

$$\underline{e}_1 . \underline{e}_3 \times \underline{e}_4 = 0 = \mu \, \underline{e}_1 . (\underline{e}_3 \times \underline{e}_2) = -\mu \, [\underline{e}_1, \underline{e}_2, \underline{e}_3]$$

and

$$\underline{e}_2 . \underline{e}_3 \times \underline{e}_4 = 0 = \lambda \, \underline{e}_2 . (\underline{e}_3 \times \underline{e}_1) = \lambda [\underline{e}_1, \underline{e}_2 \underline{e}_3]$$

Thus either λ, $\mu = 0$ which implies $\underline{e}_4 = \nu \underline{e}_3$ which is absurd

or

$$[\underline{e}_1, \underline{e}_2, \underline{e}_3] = 0 \tag{4.13}$$

Similarly

$$[\underline{e}_1, \underline{e}_2, \underline{e}_4] = 0 \tag{4.14}$$

Note that the direct proof, dotting (4.7) with $\underline{e}_1 \times \underline{e}_2$ cannot be used as it is difficult to assert that

$$\begin{bmatrix} b_4 \\ a_4 \\ c_4 \end{bmatrix} = \pm \begin{bmatrix} a_3 \\ -b_3 \\ c_3 \end{bmatrix}$$

is an unacceptable solution.

Returning to the parent equation (4.7) and squaring each row and adding gives

$$[a_1 b_4 + b_1 a_4 + c_1 c_4] \underline{e}_1 . \underline{e}_4 = [-b_2 a_3 - a_2 b_3 + c_2 c_3] \underline{e}_2 . \underline{e}_3 \tag{4.15}$$

From (4.3) and (4.6)

$$\underline{e}_1 \cdot \underline{e}_4 = A_1 F_A \cdot F_A^T M_1^T L^T A_4^T = A_1 [I - 2A_1^T A_1] GRA_4^T$$

But $A_1[I - 2A_1^T A_1] = -A_1$ since $A_1 A_1^T = 1$

Therefore

$$\underline{e}_1 \cdot \underline{e}_4 = -A_1 GRA_4^T = -[a_1 b_4 + b_1 a_4 + c_1 c_4] \tag{4.16}$$

Similarly

$$\underline{e}_2 \cdot \underline{e}_3 = A_2 LF_A \cdot F_A^T RM_2^T LA_3^T$$

$$= A_2 [I - 2A_2^T A_2] GLA_3^T$$

$$= -A_2 GLA_3^T = b_2 a_3 + a_2 b_3 - c_2 c_3 \tag{4.17}$$

(4.15) is thus reduced to

$$[\underline{e}_1 \cdot \underline{e}_4]^2 = [\underline{e}_2 \cdot \underline{e}_3]^2 \tag{4.18}$$

So

$$\underline{e}_1 \cdot \underline{e}_4 = \pm \underline{e}_2 \cdot \underline{e}_3 \tag{4.19}$$

We now write $\underline{e}_1 \cdot \underline{e}_2 = \cos\theta$ and (4.20)

$$\underline{e}_3 = [\underline{e}_1 + p\underline{e}_2]/h \tag{4.21}$$

$$\underline{e}_4 = [\underline{e}_2 - q\underline{e}_1]/j \tag{4.22}$$

where $h^2 = 1 + p^2 + 2p\cos\theta$ and $j^2 = 1 + q^2 - 2q\cos\theta$ (4.23)

so that $\underline{e}_3$ and $\underline{e}_4$ are unit vectors.

Because $\underline{e}_3$ is roughly aligned with $\underline{e}_1$ and $\underline{e}_4$ is roughly aligned with $\underline{e}_2$, h and j are both positive quantities. Relation (4.18) gives

$$(p + \cos\theta)^2/h^2 = (\cos\theta - q)^2/j^2$$

which may be written

$$\frac{h^2 - \sin^2\theta}{h^2} = \frac{j^2 - \sin^2\theta}{j^2}$$

and since θ is close to 90°, $\sin\theta \neq 0$, we have $h = j$. We can therefore write

$$\underline{e}_3 = (\underline{e}_1 + p\underline{e}_2)/h \tag{4.24}$$

$$\underline{e}_4 = (\underline{e}_2 - q\underline{e}_1)/h \tag{4.25}$$

We now have

$$\underline{e}_1 \times \underline{e}_3 = (\frac{p}{h})\ \underline{e}_1 \times \underline{e}_2 \tag{4.26}$$

$$\underline{e}_1 \times \underline{e}_4 = (\frac{1}{h})\ \underline{e}_1 \times \underline{e}_2 \tag{4.27}$$

$$\underline{e}_2 \times \underline{e}_3 = (\frac{-1}{h})\ \underline{e}_1 \times \underline{e}_2 \tag{4.28}$$

$$\underline{e}_2 \times \underline{e}_4 = (\frac{q}{h})\ \underline{e}_1 \times \underline{e}_2 \tag{4.29}$$

$$\underline{e}_3 \times \underline{e}_4 = \left[\frac{1 + pq}{h^2}\right] \underline{e}_1 \times \underline{e}_2 \tag{4.30}$$

(4.19) gives either $q = -p$ or $q = p + 2\underline{e}_1.\underline{e}_2$ which we may write as $q = p + 2\cos\theta$.

To decide which of these two values for q is correct, we appeal again to the main equation (4.7). Post-cross with $\underline{e}_3$, then gives

$$\begin{bmatrix} b_4 \\ a_4 \\ c_4 \end{bmatrix} \frac{(1 + pq)}{h^2} = \begin{bmatrix} a_1 \\ b_1 \\ c_1 \end{bmatrix} \frac{p}{h} + \begin{bmatrix} -b_2 \\ a_2 \\ c_2 \end{bmatrix} \frac{1}{h} \tag{4.31}$$

Squaring each row and adding, noting that $\underline{e}_1.\underline{e}_2 = \cos\theta = b_1a_2 - a_1b_2 + c_1c_2$ and that $h^2 = 1 + p^2 + 2p\cos\theta$, we obtain

$$(1 + pq)^2/h^4 = 1$$

i.e.
$$1 + pq = \pm h^2 \tag{4.32}$$

Since a_4 in (4.31) must have the same sign as a_2 for all simple patches (both approximate to + 1) we can eliminate the negative sign and write

$$1 + pq = h^2 = 1 + p^2 + 2p\cos\theta \tag{4.33}$$

giving $q = p + 2\cos\theta$.

Rearranging (4.31) gives

$$h \begin{bmatrix} a_4 \\ b_4 \\ c_4 \end{bmatrix} = p \begin{bmatrix} b_1 \\ a_1 \\ c_1 \end{bmatrix} + \begin{bmatrix} a_2 \\ -b_2 \\ c_2 \end{bmatrix} \quad (4.34)$$

Using a similar procedure, post-crossing (4.7) with $\underline{e}_4$ gives

$$h \begin{bmatrix} a_3 \\ b_3 \\ c_3 \end{bmatrix} = \begin{bmatrix} a_1 \\ -b_1 \\ c_1 \end{bmatrix} + q \begin{bmatrix} b_2 \\ a_2 \\ -c_2 \end{bmatrix} \quad (4.35)$$

If (4.34) and (4.35) together with (4.24) and (4.25) are substituted in (4.7), each row of the latter is an identity revealing that there are no hidden additional constraints.

5. THE SPECIAL CASE OF A DUPIN CYCLIDE

If the plane curves are <u>arcs of circles</u> the $\underline{e}$ vectors are directed along the chords. $\underline{e}_3$ and $\underline{e}_4$ now lie in the plane of $\underline{e}_1$ and $\underline{e}_2$ namely the E plane. If $\underline{e}_3$ and $\underline{e}_4$ are produced, they intersect at corner C so that the position matching equation becomes a relation between chord lengths:

$$\ell_1 \underline{e}_1 + \ell_4 \underline{e}_4 = \ell_2 \underline{e}_2 + \ell_3 \underline{e}_3 \quad (5.1)$$

Post-crossing with $\underline{e}_4$ and $\underline{e}_3$ in turn gives

$$h\ell_3 = \ell_1 - q\ell_2 \quad (5.2)$$

$$h\ell_4 = \ell_2 + p\ell_1 \quad (5.3)$$

Since K for each curve is now constant, and ϕ is constant, both the geodesic curvature g and the normal curvature of each side are constant. The chord length is $\ell = 2R \sin \frac{\psi}{2}$, so $K = \frac{2 \sin \frac{\psi}{2}}{\ell}$.

Multiplying by $\sin\phi$ and $\cos\phi$ in turn gives

$$g = \frac{2b}{\ell} \quad (5.4)$$

$$n = \frac{2c}{\ell} \quad (5.5)$$

It is useful to have a letter for $\frac{2a}{\ell}$ so we define

$$f = \frac{2a}{\ell} \quad (5.6)$$

Note that

$$f^2 + q^2 + n^2 = \frac{4}{\ell^2} \text{ since } a^2 + b^2 + c^2 = 1 \qquad (5.7)$$

We may now find expressions for the g's and n's of sides 3 and 4 of a cyclide patch in terms of our parameters p and q as

$$g_3 = \frac{2b_3}{\ell_3} = \frac{2(qa_2 - b_1)}{\ell_1 - q\ell_2} = -\Gamma g_1 - (1 - \Gamma)f_2 \qquad (5.8)$$

$$n_3 = \frac{2c_3}{\ell_3} = \frac{2(c_1 - qc_2)}{\ell_1 - q\ell_2} = \Gamma n_1 + (1 - \Gamma)n_2 \qquad (5.9)$$

$$g_4 = \frac{2b_4}{\ell_4} = \frac{2(-b_2 + pa_1)}{\ell_2 + p\ell_1} = -\Delta g_2 + (1 - \Delta)f_1 \qquad (5.10)$$

$$n_4 = \frac{2c_4}{\ell_4} = \frac{2(c_2 + pc_1)}{\ell_2 + p\ell_1} = \Delta n_2 + (1 - \Delta)n_1 \qquad (5.11)$$

where

$$\Gamma = \frac{\ell_1}{\ell_1 - q\ell_2} \;;\quad 1 - \Gamma = \frac{-q\ell_2}{\ell_1 - q\ell_2} \qquad (5.12)$$

$$\Delta = \frac{\ell_1}{\ell_2 + p\ell_1} \;;\quad 1 - \Delta = \frac{p\ell_1}{\ell_2 + p\ell_1} \qquad (5.13)$$

Thus n_3 and n_4 are a blend of only n_1 and n_2 but g_3 and g_4 are a more complicated blend.

Note that $n_1 = n_2$ gives $n_3 = n_4 = n_1 = n_2$

If we attempt to set $n_3 = n_4$ then

$$\Gamma n_1 + (1 - \Gamma)n_2 = \Delta n_2 + (1 - \Delta)n_1$$

i.e.

$$(\Gamma + \Delta - 1)(n_1 - n_2) = 0$$

But $\Gamma + \Delta - 1 = \ell_1\ell_2 h^2/(\ell_1 - q\ell_2)(\ell_2 + p\ell_1) \neq 0$ so $n_1 = n_2$.

A cyclidal patch is either a piece of a sphere ($n_1 = n_2$) or else cannot contain a spherical point anywhere ($n_1 \neq n_2$).

6. THE CYCLIC QUADRILATERAL OF A CYCLIDAL PATCH

The relation $\underline{e}_1 \cdot \underline{e}_4 = -\underline{e}_2 \cdot \underline{e}_3$ implies that the angle between $\underline{e}_2$ and $\underline{e}_3$ and the angle between $\underline{e}_1$ and $\underline{e}_4$ add up to 180°. The angles shown in Figure 5 are therefore equal. The chords of a cyclide therefore not only form a plane quadrilateral in the E plane but this quadrilateral is cyclic.

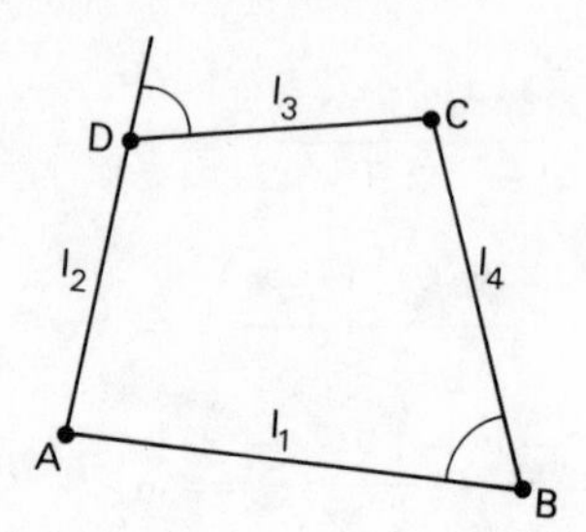

Fig. 5 The chords of a cyclide form a cyclic quadrilateral

The diagonal vectors $\vec{BD}$ and $\vec{AC}$ are given by

$$\underline{\ell}_5 = \vec{BD} = \underline{\ell}_2 - \underline{\ell}_1 \tag{6.1}$$

$$\underline{\ell}_6 = \vec{AC} = \underline{\ell}_2 + \underline{\ell}_3 = (\underline{\ell}_1/\Gamma) + (\underline{\ell}_2/\Delta) \tag{6.2}$$

$$\ell_6 = \ell_5/h \tag{6.3}$$

7. THE SHAPE PARAMETER P FOR A CYCLIDE PATCH

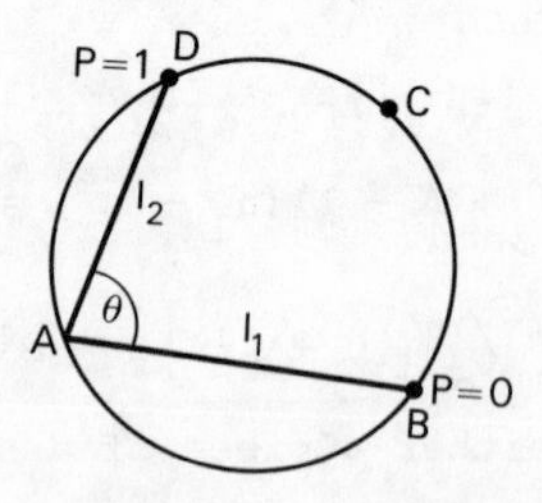

Fig. 6 Point C lies on Circle

The corner C of a cyclide patch lies on the circumcircle of a triangle ABD as shown in Figure 6 (which is entirely determined by the given sides 1 and 2). If the diagonal AC lies in the angle DAB the patch is open. Once C progresses past D or

B the patch becomes crossed. We introduce a parameter P that is 0 when C is at B and 1 when C is at D. This parameter has been designed so that both P → p,q and p,q → P are simple relationships.

Let
$$\ell_5 = [\ell_1^2 + \ell_2^2 - 2\ell_1\ell_2\cos\theta]^{\frac{1}{2}} = BD \tag{7.1}$$

$$\ell_7 = [\ell_1^2 + \ell_2^2 + 2\ell_1\ell_2\cos\theta]^{\frac{1}{2}} \tag{7.2}$$

Then
$$\ell_4 = P\ell_1\ell_5/L \tag{7.3}$$

$$\ell_3 = (1-P)\ell_2\ell_5/L \tag{7.4}$$

where
$$L = [\ell_1^2P^2 + \ell_2^2(1 - P)^2 + 2P(1 - P)\ell_1\ell_2\cos\theta]^{\frac{1}{2}} \tag{7.5}$$

Then
$$p = [P\ell_5^2 - \ell_2^2]/\ell_1\ell_2 \tag{7.6}$$

$$q = [\ell_1^2 - (1 - P)\ell_5^2]/\ell_1\ell_2 \tag{7.7}$$

$$h = \sqrt{1 + pq} = \ell_5L/\ell_1\ell_2 \tag{7.8}$$

Inversely
$$P = \frac{\ell_2(\ell_2 + p\ell_1)}{\ell_5^2} \tag{7.9}$$

$$1 - P = \frac{\ell_1(\ell_1 - q\ell_2)}{\ell_5^2} \tag{7.10}$$

This gives the simple result that

$$AC = \ell_6 = \ell_1\ell_2/L \tag{7.11}$$

Note that Ptolemy's Theorem

$$\ell_1\ell_3 + \ell_2\ell_4 = \ell_5\ell_6$$

is now easily established.

The vector
$$\underline{\ell}_6 = \vec{AC} = \frac{(1 - P)\ell_2^2\underline{\ell}_1 + P\ell_1^2\underline{\ell}_2}{L^2} \tag{7.12}$$

Also
$$\Gamma = \ell_1^2/\ell_5^2(1 - P) \tag{7.13}$$

$$\Delta = \ell_2^2/\ell_5^2P \tag{7.14}$$

If $r = \ell_2/\ell_1$, the aspect ratio, then $p = \frac{r^2}{r^2+1}$ is the parameter that leads to the condition of equal stretch: $\ell_3 = R\ell_1$, $\ell_4 = R\ell_2$ where $R = \ell_5/\ell_7$.

For the equal stretch patch

$$\ell_3 = \ell_1\ell_5/\ell_7 \tag{7.15}$$

$$\ell_4 = \ell_2\ell_5/\ell_7 \tag{7.16}$$

$$\ell_6 = (\ell_1^2 + \ell_2^2)/\ell_7 \tag{7.17}$$

$$\underline{\ell}_6 = \frac{\ell_1^2 + \ell_2^2}{\ell_7^2}[\underline{\ell}_1 + \underline{\ell}_2] \tag{7.18}$$

Interestingly for this case

$$n_3 + n_4 = n_1 + n_2 \tag{7.19}$$

If θ is acute n_3 and n_4 lie outside n_1 and n_2.
If θ is obtuse n_3 and n_4 lie inside n_1 and n_2.

8. THE INTERIOR OF A CYCLIDAL PATCH

The general interior point is given by

$$\underline{r}(u,v) = \frac{u[\ell_1 - vq(v)\ell_2]\ \underline{e}_1(u) + v[\ell_2 + up(u)\ell_1]\underline{e}_2(v)}{1 + uvp(u)q(v)} \tag{8.1}$$

where

$$\underline{e}_1(u) = u\underline{e}_1 + n_1\ \underline{t}_1 \tag{8.2}$$

$$\underline{e}_2(v) = v\underline{e}_2 + v_1\ \underline{t}_2 \tag{8.3}$$

$$p(u) = up + 2u_1b_2 = up + u_1g_2\ell_2 \tag{8.4}$$

$$q(v) = vq + 2v_1b_1 = vq + v_1g_1\ell_1 \tag{8.5}$$

$$u_1 = [u^2a_1^2 + 1 - u^2]^{\frac{1}{2}} - ua_1 \tag{8.6}$$

$$v_1 = [v^2a_2^2 + 1 - v^2]^{\frac{1}{2}} - va_2 \tag{8.7}$$

A more symmetric view of u and u_1, and v and v_1 is given by sub-parameters λ and μ

$$\left.\begin{aligned} u &= \sin[\tfrac{1}{2}\lambda\psi_1]/\sin(\tfrac{1}{2}\psi_1) \\ u_1 &= \sin[\tfrac{1}{2}(1-\lambda)\psi_1]/\sin(\tfrac{1}{2}\psi_1) \\ v &= \sin[\tfrac{1}{2}\mu\psi_2]/\sin(\tfrac{1}{2}\psi_2) \\ v_1 &= \sin[\tfrac{1}{2}(1-\mu)\psi_2]/\sin(\tfrac{1}{2}\psi_2) \end{aligned}\right\} \quad (8.8)$$

The proof of this result is involved, and must be left to a separate paper.

Lines of constant u and constant v are lines of curvature. Other results are

$$p(u,v) = vp(u) \quad (8.9)$$

$$q(u,v) = uq(v) \quad (8.10)$$

$$A_1(u) = [ua_1 + u_1, ub_1, uc_1] \quad (8.11)$$

$$A_2(v) = [va_2 + v_1, vb_2, vc_2] \quad (8.12)$$

9. CONCLUDING SECTION

The preliminary ideas on frame and position matching were worked out in discussions between myself and Mrs. F.J.M. Nunn. Ralph Martin then joined my research team as a research student and solved the frame matching equation for plane lines of curvature directly from the rotation matrices. Only later was the Householder's form discovered and I then solved the equation by splitting it into symmetric and skew symmetric parts. This stripped out the redundancy because the symmetric parts could be expressed in terms of the skew symmetric parts (which have zeros on the diagonal). The symmetric parts only carried additional sign information. Dr. Martin then introduced the $\underline{e}$ vectors and proved that the corners of a cyclide are co-planar. I showed that the chord quadrilateral is cyclic. (These facts can be proved by inversion geometry and were established by earlier writers on the cyclide). Dr. Martin [1] has refined the methods still further by introducing quaternions to express the rotation of frames. This avoids the redundancy inherent in orthogonal matrices.

The solution I present here is the best that I can devise using the simplest mathematical techniques - just matrices and vectors - because this is the common language of engineers. Quaternions, Differential Forms, Frame Fields etc. are not yet on our undergraduate menu.

John du Pont [2] has solved the frame matching equation for helical lines of curvature using quaternions, but such a patch has non-helical lines of curvature in the interior. It is possible that spherical lines of curvature will avoid this difficulty and yield a patch that has boundary curves that are usefully non-planar. A spherical line of curvature has constant n but unrestricted g. Compare this with a circular line of curvature that bounds a cyclidal patch which has constant n and g.

T. Sharrock has invented a most elegant method for making models of an assembly of cyclic quadrilaterals. This has enabled us to visualise surfaces composed of cyclidal patches.

The cyclidal patch has other qualities that suggest that it is a real contender for a new design technique. M. Sabin [3] has extolled its virtues for calculating intersections and J. de Pont [2] has made extensive investigations.

I have made no mention of the knotty problem of existence. Does a surface patch created via frame matching and position matching satisfy the Codazzi-Mainardi equations? Martin [1] has shown that the requirements are closely linked - an elegant piece of work locked up in his Ph.D thesis that deserves a more public presentation.

We do not need to get too concerned about surface existence for the cyclide. It has such a long history with two splendid papers by Maxwell [4] and Cayley [5] expounding the properties of complete cyclide surfaces. My discovery of a parametric formula for $\underline{r}(u,v)$ for the interior of a cyclide patch clinches the matter. It is always the acid test for surface existence.

ACKNOWLEDGEMENTS

I am indebted to the IMA typist (Miss Pamela Irving) who has kindly prepared both my manuscripts for the press; and to the Editor, Dr. Gregory.

REFERENCES

[1] Martin, R.R., Principal Patches for Computational Geometry, *Ph.D. Thesis*, Cambridge University, 1982

[2] De Pont, J., Essays on the Cyclide Patch, *Ph.D. Thesis*, Cambridge University, 1984

[3] Sabin, M.A., Private Communication, Cambridge, 1983

[4] Maxwell, J.C., On the Cyclide, Quarterly Journal of Pure and Applied Mathematics, Vol. 9, 1868

[5] Cayley, A., On the Cyclide, Quarterly Journal of Pure and Applied Mathematics, Vol. 12, 1873.

CYCLIDE SURFACES IN COMPUTER AIDED DESIGN

R.R. Martin
(University College Cardiff)

J. de Pont, T.J. Sharrock
(Cambridge University Engineering Department)

1. INTRODUCTION

In a previous paper in these proceedings, Nutbourne [9] discussed a method of creating principal patches, a class of surface patch based on lines of curvature. The method requires the simultaneous solution of two sets of equations - the frame matching equations, and the position matching equations. While it is possible to solve the former for quite a wide range of patch boundary curves, the same does not seem to be true for the latter. However, in the particularly simple case where the patch edges are circular arcs, a solution can be obtained. See also Martin [8].

It turns out that surfaces all of whose lines of curvature are circular arcs were well known to differential geometers. They were originally discovered by Dupin [3], and are now known as Dupin's cyclides. A discussion of some of their properties which were known classically may be found in Hilbert and Cohn-Vosson [6] and Darboux [2]. Note however, that we shall be using small pieces of such surfaces, bounded by lines of curvature, to form composite surfaces for use in computer aided design, rather than considering whole cyclides from an analytic point of view. Thus, most of the properties we shall present here are new, and specifically useful for surface creation using cyclide patches.

2. GENERAL CYCLIDES AND SPECIAL CASES

Spheres and toruses both have circular lines of curvature, and if we allow straight lines to be considered as limiting cases of circles with infinite radius, then we can also include circular cylinders and cones, and finally the plane, in the class of Dupin's cyclides, as special cases. This is important from an engineering point of view, as the above surfaces are often traditionally used, and many other patch systems can only provide an approximate representation of these shapes.

More generally, the easiest way of imagining a cyclide is to think of it as being like a torus, but whose radius varies as we go round. If this variation goes to zero and then increases again, the result is a horned cyclide. Various examples of cyclides are shown in Figure 1. Note however that the overall shape of cyclides is relatively unimportant, as we shall only be using small patches extracted from them, to form composite surfaces.

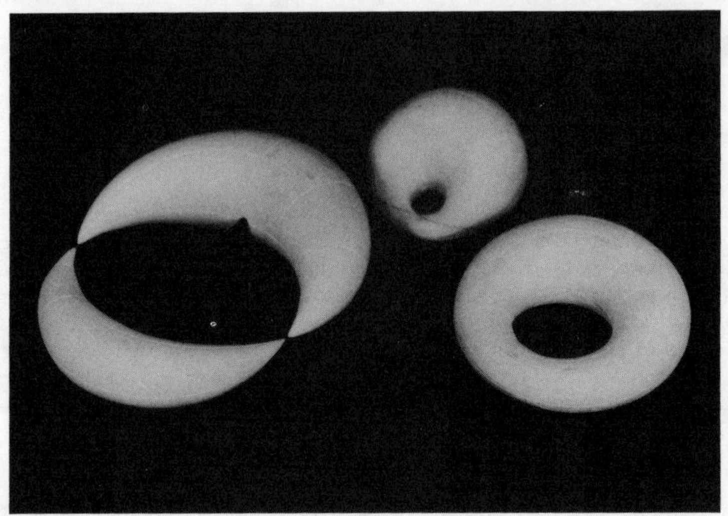

3. CYCLIC QUADRILATERAL PROPERTY

One important property of cyclide patches is that the four corners of the patch lie on a single circle, and are thus the vertices of a cyclic quadrilateral. It can be seen that this imposes a severe restriction on the choice of corners for a cyclide patch. In simple terms, if we take two adjacent sides of the patch as being given, then the fourth corner of the patch is constrained to move around on the circle defined by the three given corners, and so we only have one degree of freedom (the angular position of the fourth corner on the circle) with which we can alter the overall shape of the patch. Thus, if we are building up a composite surface from an array of patches, starting with two piecewise circular arc lines of curvature, we will only have one free choice to make per patch, which is rather restrictive. Ways round this problem will be discussed later.

4. DESIGN STRATEGY

The basic method for defining a cyclide patch assumes that two adjacent sides of the patch are already fully specified, leaving one free parameter to adjust the shape of the patch. This affects the way in which complete surfaces are constructed. Two primary lines of curvature are chosen, mutually perpendicular and composed of arcs of circles. These may be constructed, for example, as biarc splines, Bolton, [1]. If the surface has a line of symmetry, this will usually be chosen, and the other primary line chosen to bisect, approximately, the surface. A surface normal at the intersection is also specified. Patches are then propagated out from the four corners of the mesh, each patch nestling against two previously constructed patches or against a primary line (Figure 2).

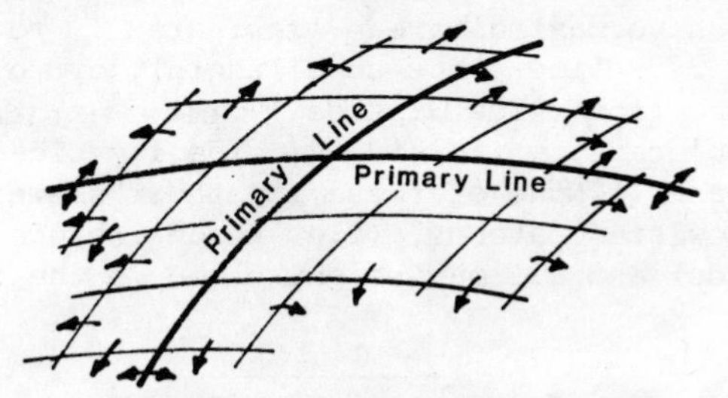

Fig. 2. Propagation of patches

This still leaves one degree of freedom per patch, and it is useful to have an automatic procedure to create a surface which can then be modified until the designer is satisfied. Unless the internal modification techniques described below are used, changing a patch will alter all patches downstream of it. However, if the free parameter of each patch is used to satisfy an external condition, the effect of these changes will be minimised.

A number of different conditions are available, which fall into two classes: those which depend only on the underlying cyclic quadrilateral, and those which make use of curvature information. If only the former are used the surface then has a fixed 'skeleton' (the faceted surface made up from the cyclic quadrilaterals on which the patches sit), which does not vary as the initial surface normal is altered. The fairness of the surface varies, but it continues to pass through all the vertices of the skeleton. (See Figure 3). If conditions of the latter sort are satisfied, the skeleton itself varies as the normal is shifted.

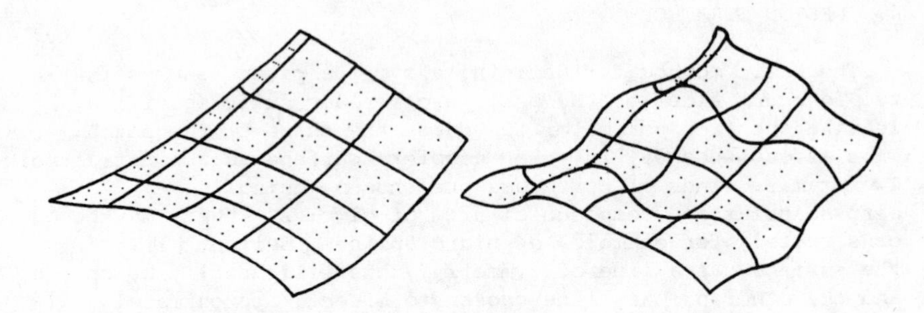

Fig. 3 Two surfaces with the same skeleton

Various conditions of the first type are illustrated in Figure 4. Given the two leading chords, the cyclic quadrilateral can be chosen to have maximal area ('maxi-area'), to be triangular (either side 3 or side 4 has zero length), or to have its sides in proportion (side 3: side 1 = side 4: side 2 is the 'equi-stretch' case, while side 4: side 1 = side 3: side 2 is the 'kite' case). None of these, or their skewed relatives, can produce twisted patches, (i.e. which include the horn of a horned cyclide) and all may be expressed in the form

$$P = \frac{1}{1 + r^{-a} e^{-b}} \tag{4.1}$$

where P is the shape parameter for the patch, r is the ratio side 2: side 1, a is the 'power' of the patch, and b its skewness. For example, an equi-stretch patch has power 2.0 and skewness 0.0, while a maxi-area has power 1.0 and skewness 0.0. In addition, side 3 or side 4 may be forced parallel to side 1 or side 2 respectively, forming a 'trapezoidal' patch, which is needed to represent a surface of revolution, or the free diagonal may be made into a diameter. These conditions, however, may result in twisted patches.

If curvature information is used, it is possible to choose any one of the normal or geodesic curvatures of sides 3 & 4, within limits, or to freely choose the Gaussian curvature of the free corner. It is also possible, numerically, to optimise some function, for example to minimise the mean angle swung through by the patch boundaries. This is in some sense equivalent to splining, as it attempts to minimise oscillation of the surface.

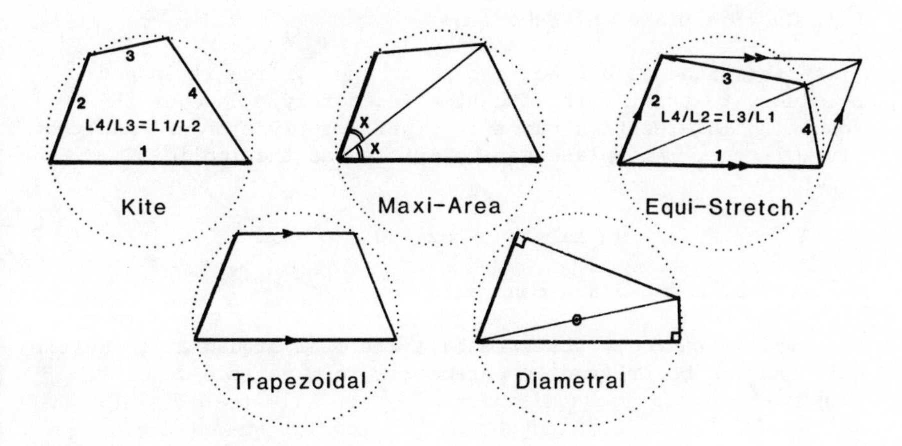

Fig. 4. Types of cyclic quadrilateral

5. SURFACE INTERSECTIONS

A fundamental problem in the application of a surface patch to design is the determination of intersection curves. If the two surfaces to be intersected are both described by parametric equations (the usual situation for surface patches) the intersection curve can be described by three simultaneous equations in four variables.

$$\underline{r}_1(u_1,v_1) - \underline{r}_2(u_2,v_2) = 0 \qquad (5.1)$$

In principle we can eliminate two of the variables to obtain an implicit equation in two variables. However as the equations are non-linear this is not a trivial task.

If, however, one of the surfaces can be described by an implicit equation the elimination is achieved automatically by substituting the parametric equation of one surface into the implicit equation of the other. To this end Sederberg et al [11] developed a method for determining the implicit equation of a bicubic patch. In this case the surface has degree 18 and the resulting intersection curve degree 108. Fortunately the equivalent cyclide patch results are much more manageable. Three intersections will be considered; cyclide-plane, cyclide-cyclide and cyclide-quadric surface. Having found an implicit equation for the intersection curve it is usual to evaluate points on it using a marching algorithm. As the cyclide patch is based on its principal lines of curvature, the step sizes for this algorithm can be based on exact curvatures rather than numerical approximations.

5.1. Cyclide-plane intersections

As the plane is a special case of the cyclide it is not necessary to treat this situation separately. However the resulting intersection curve is significantly simpler than the general case. Any plane can be written as the implicit equation,

$$ax + by + cz + d = 0 \tag{5.2}$$

where a, b, c and d are constants.

The cyclide has been shown to be a rational biquadratic (Martin, [8]) and can be written in parametric form as (see de Pont, [10])

$$\underline{r}(u,v) = \frac{1}{A_1A_2+B_1B_2}\begin{bmatrix} a_1u\,(A_2\ell_1-B_2\ell_2) - b_2v^2(A_1\ell_2+B_1\ell_1) \\ b_1u^2(A_2\ell_1-B_2\ell_2) + a_2v\,(A_1\ell_2+B_1\ell_1) \\ c_1u^2(A_2\ell_1-B_2\ell_2) + c_2v^2(A_1\ell_2+B_1\ell_1) \end{bmatrix}^T \tag{5.3}$$

where $A_1 = a_1^2+(b_1^2+c_1^2)u^2$ $\quad A_2 = a_2^2+(b_2^2+c_2^2)v^2$

$B_1 = (pu + 2b_2a_1\ (1-u))u$ $\quad B_2 = (qv + 2b_1a_2(1-v))v$

and $\ell_i(a_i, b_i, c_i)$ is the chord vector of side i of the patch.

Provided we write the plane equation in terms of the same coordinate axes as the cyclide patch equation we can substitute (5.3) into (5.2) to get the equation of the intersection curve. This has the following form:

$$\alpha uv^2 + \beta u^2v + \gamma uv^2 + \delta u^2 + \varepsilon uv + \lambda v^2 + \eta u + \mu v + \omega = 0 \tag{5.4}$$

where α, β, γ, δ, ε, λ, η, μ and ω are constant coefficients.

Fixing u (or v) leaves us with a quadratic equation in v (or u). Furthermore, regarding the equation as a quadratic makes the discriminant a quartic expression. Finding the roots of this enables us to find the asymptotes and double points of the curve. Thus this intersection curve has a simple form and can be completely evaluated analytically.

5.2 Cyclide-cyclide intersections

As the cyclide patch lies on a Dupin cyclide we know that it has an implicit equation. Provided we can find a method of converting from parametric form to implicit form we have a simple method of determining the intersection of two cyclide patches.

A Dupin cyclide has two planes of symmetry which, in the notation used by Forsythe [5], intersect along the x axis. The surface cuts the x axis at four points. By extending the parameters of the cyclide patch we find the four points which lie on the intersection of the two planes of symmetry. Comparing these with the values for a general Dupin cyclide we can evaluate the constants of the implicit equation and determine the orientation of the underlying surface. Having established an implicit equation for one patch we can substitute the parametric equation of the other to get an equation for the intersection curve. This turns out to be an eighth order equation of the form,

$$[u^4, u^3, u^2, u, 1]\,[G] \begin{bmatrix} v^4 \\ v^3 \\ v^2 \\ v \\ 1 \end{bmatrix} = 0 \qquad (5.5)$$

or

$$\underline{u}^T\,[G]\,\underline{v} = 0$$

Fixing u (or v) and finding v (or u) involves solving a quartic equation which can be done analytically.

To check that points on the intersection curve lie within the patch boundaries we must convert from position values to parameter values. Forsythe's parametric equations for the complete Dupin cycle can be manipulated to allow this to be done.

5.3 Cyclide-quadric intersections

Sculptured surface methods are now being integrated into solid modelling systems, Jared et al [7]. To achieve this the surfaces must be able to be intersected with any other object the solid modeller can create. As most solid modelling systems include quadric surfaces there is a need to be able to do cyclide-quadric intersections. Note that some of the quadrics (cones, spheres, cylinders) are also cyclides and could be handled by the cyclide-cyclide intersection facility. A quadric surface is described by an implicit equation and so we can determine the equation of the intersection curve by substituting the parametric equations of the patch. The resulting equation has a form analogous to (5.5) and can be solved in the same way.

$$\underline{u}^T [Q] \underline{v} = 0 \tag{5.6}$$

5.4 Step Size Determination

For plotting or machining an intersection curve it is usual to approximate it by a sequence of straight line segments. If we set limits on the maximum deviation δ between the actual curve and the relevant straight line segment then for small steps it can be shown (Faux et al, [4]) that:

$$L^2 = 4\delta(2\rho - \delta) \tag{5.7}$$

where L is the step length,

and ρ is the radius of curvature.

For most surface patches it is necessary to use a numerical method to estimate ρ. However with the cyclide patch the parameter lines are both lines of curvature and circular arcs. Thus it is a simple matter to calculate the normal curvature at any point in any direction. The normal curvature of the plane is, of course, zero and for a quadric surface it can be calculated. From the normal curvature on the two surfaces the curvature of the intersection curve can be determined. Thus the step length can be derived from the exact curvature at a point. Having found a step length, $\delta s = L$, this needs to be converted to a parameter step, δu or δv. The parametric equation of the patch can be differentiated to give $\delta s/\delta u$ and $\delta s/\delta v$. As the direction of the intersection curve tangent is known relative to the parameter line tangents, the step length L can be converted to a corresponding increment in the parameter u or v.

6. BLENDING

The standard formulation of the cyclide patch is to regard two adjacent sides as given and to choose a single parameter to complete the patch. If we consider the problem of joining two cyclide patch surfaces together using a blend composed of cyclide patches we see that the problem is one of fitting cyclide patches when three boundaries are given. Two methods for achieving this have been developed; the two patch blend and the three patch blend. In figure 5 we illustrate a row of patches across the blend.

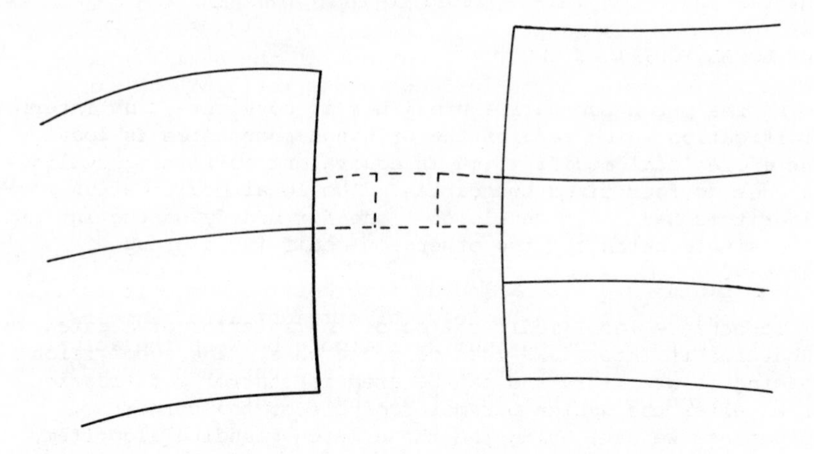

Fig. 5. First row of patches blending two surfaces

A cyclide patch boundary is specified by three parameters, κ, the curvature, ϕ, the angle between the curve normal and the surface normal, and s, the arc length. If we consider the blend as a sequence of patches generated from left to right we have to meet three conditions to satisfy the right hand boundary and thus we expect to require three patches across the blend. However, if we allow the arc length for the right hand boundary to be variable, we need only satisfy two conditions and may be able to blend with rows of only two patches across the blend.

Using the cyclic quadrilateral property of the chords of the cyclide patch we can determine the patch parameters for both the two and the three patch blends. To commence the blending operation we choose a point on each of the surfaces and join them by two or three circular arcs depending on the type of blend to be used. We then apply the blending algorithm generating rows of patches in both directions away from this joining curve. Each patch vertex on the edges to be joined blends to some point on the edge of the other surface. With the two patch method the blending algorithm automatically fixes this point. There is no choice. The three patch algorithm allows these points to be chosen arbitrarily. The most obvious result of this difference is that the three patch blend can always include the whole edge of both surfaces while the two patch blend cannot. There is nothing in the formulation of either blend to prevent the formation of twisted patches although it is easier to avoid them with the three patch method.

Both of these algorithms have been tested to see that they work but they have not been applied to real blends.

7. LOCAL MODIFICATIONS

As the patch boundaries are lines of curvature, any interior modification which retains the original boundaries is local. Therefore local modification is equivalent to fitting cyclide patches to four fixed boundaries. Two local modification algorithms have been developed; one for modifying the interior of a single patch and the other for modifying a group of patches.

To achieve local modification of the interior of a patch we subdivide it into a 3x3 mesh of subpatches. The subdivision spacing is arbitrary and can be used to control modification. If we alter the design parameter of one of the corner subpatches we can, using the three patch blending algorithm, determine the patch parameters for each of the other corner subpatches such that the subpatches along the boundary remain cyclide. To ensure that this algorithm is consistent we must check that the parameter of the subpatch diagonally opposite the the one we control does not depend on which route was taken in calculating it and that the centre subpatch also remains cyclide. As any cyclide can be inverted to either a torus, a cone or a cylinder we need only to prove this consistency for for these special cases.

By considering the number of conditions to be met in order to arbitrarily specify the third and fourth boundaries of a group of patches and, assuming one independent parameter for each patch, we expect the minimum configurations for which local modifications might be possible to be 4x10, 5x7 and 6x6. These are when the number of parameters exceeds the number of conditions. An algorithm has been developed which takes a 6x6 mesh of patches and allows the topological midpoint to be moved arbitrarily. An iterative scheme based on the three patch blend is used to determine all the changes to the patches. Although the scheme can be made to be relatively stable only modest modifications are possible without the formation of twisted patches.

8. THE CYCLIDE-Q PATCH

One of the major limitations of the cyclide patch is that if we regard it conventionally as having two boundaries fixed we have only one parameter to design with. Several approaches to increasing the flexibility of the patch have been tried of which the cyclide-Q patch is the most promising.

Essentially the concept is that by considering four patches simultaneously as a single patch we can increase the complexity of the system while retaining the simple properties of the individual patches. The idea has been used previously by Varady [12] with his double quadratic patches.

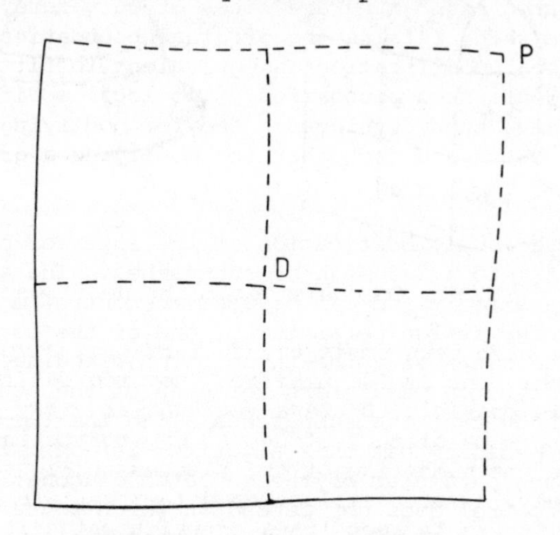

Fig. 6. Cyclide Q-patch

Consider a cyclide Q-patch with two boundaries fixed as shown in figure 6. Suppose we wish to choose the position of point P. A point is specified by three coordinates and we have four parameters so we expect to have one free parameter. Using this to design the bottom left patch we find the point D. Inverting the system about a sphere centred on D simplifies the problem to one of straight lines and planes which we can easily solve for the three remaining patch parameters. Thus the fourth corner of a cyclide Q-patch can be chosen as a design parameter with one further parameter for controlling the patch interior. This development enables the cyclide patch to be used for surface fitting and for modifications to be achieved by moving points on the surface.

9. UMBILIC POINTS

At umbilic points (points at which the principle curvatures are equal) the rectangular arrangement of lines of curvature breaks down. Two common examples, a rounded corner and an axis of revolution are shown in Figure 7. If surfaces involving umbilic points are to be modelled (and most complete surfaces include them), a way must be found of representing the nonrectangular topology.

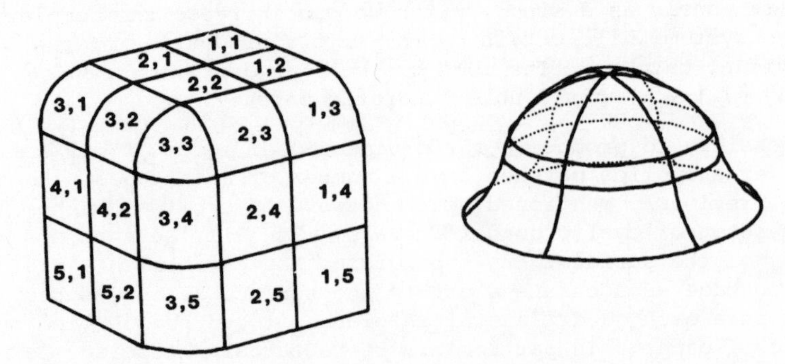

Fig. 7. Commonly occurring umbilics

Two methods have been employed, so far. For simple cases, involving only one or two umbilics, not too far from the outer edges, it is possible to use a rectangular array of patches, but allowing some of the patches to degenerate into lines of points (see the numbering of the patches in Figure 7). Alternatively, a more complicated data-structure, involving numerous pointers to keep track of which patch is adjacent to which, is constructed. This data structure requires complex housekeeping, to ensure that it is always consistent. The only cyclide which contains an umbilic is the sphere, and this surface is entirely umbilic. Thus whenever an umbilic is required a spherical patch is needed, and the required condition for this is that the patch should have equal normal curvatures for its two leading sides. It is therefore necessary to decide that an umbilic is wanted upstream of where it is actually required. This can be a significant constraint.

10. INTERACTIVE DESIGN

The design of free-form surfaces by mean of lines of curvature is an unfamiliar technique, and efficient interactive tools are still being developed. For complex surfaces, the lines of curvature are not the most natural lines on the surface, and it is essential to have some means of shape feedback to the designer. In particular a relatively small modification of a surface can greatly change the positions, and even the topology, of the lines of curvature, for example by introducing an umbilic point.

Wire-frame drawings, showing the edges of the patches, can be produced quickly at a graphics terminal, and show the broad effects of modifications. Shaded pictures are much slower and more expensive to produce, but do give a better impression of the surface - an important factor here is that significant oscillations in geodesic curvature do not necessarily result in a wavy surface, and hence the network of lines of curvature may be a pessimistic image of the fairness of the surface.

In order to get an even better impression of the surface, it is possible to produce a three-dimensional model of it, and in this area cyclide patches have a number of advantages. As has been previously mentioned, each network of cyclide patches has a skeleton of cyclic quadrilaterals, with all the vertices lying on the surface. By subdividing the patches this skeleton can be made as accurate a fit to the surface as is desired, and since each quadrilateral is plane, it can be modelled by a piece of paper. In particular, if rectangular topology is preserved, two sets of strips of paper can be drawn, each strip can be folded along straight lines, and each set then covers the surface. By making the two sets of strips lie in perpendicular directions, and then glueing corresponding quadrilaterals together, a 'rigid' model can be produced (Figure 8). Surfaces with low Gaussian curvature will however not be stiff, as by nature they are almost developable.

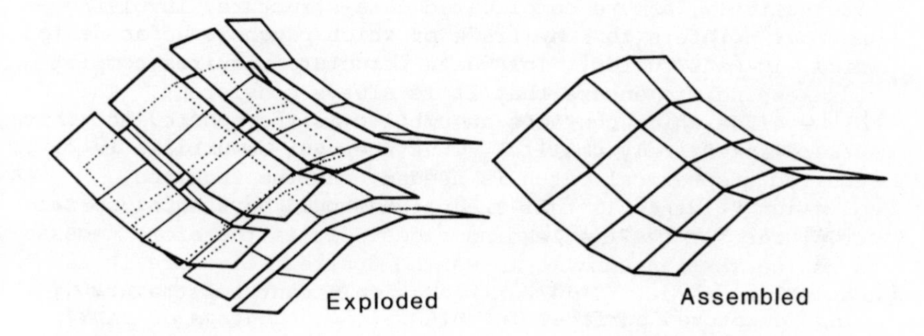

Fig. 8. Skeleton Model of a Surface

Alternatively, a Numerically-Controlled machine tool can be used to carve a model of the surface. A ball-ended cutter is used, and is driven over a surface a constant distance away from the required surface. This offset surface is particularly easy to find for a cyclide, as it is another cyclide (Martin, [8]). This means that the offset surface can be handled in exactly the same way as the given surface, and also that if, as is normal, the surface is machined along its parameter lines, these will all be arcs of circles, with great potential savings in the amount of data required to be transmitted.

A number of experiments have been done in interactive design with cyclide patches, and it has become apparent that more work needs to be done to give the systems more flexibility, and to 'train the intuition' of the designer.

ACKNOWLEDGEMENTS

The authors would like to thank A.W. Nutbourne for the original suggestion that they should investigate cyclides, and for many helpful discussions and much encouragement with this work.

REFERENCES

[1] Bolton, K.M., Biarc curves, Computer Aided Design, 7(2) pp. 89-92, 1975.

[2] Darboux, G., *Systemes orthogonaux et les coodonees curvilignes* (Second edition), Gauthier-Villars et Fils, Paris, 1910.

[3] Dupin, C., *Applications de geometrie et de mecanique,* Bachelier, Paris, 1822.

[4] Faux, I.D., Pratt, M.J., *Computational geometry for design and manufacture,* Ellis Horwood, Chichester, 1979.

[5] Forsythe, A.R., *Lectures on differential geometry of curves and surfaces,* Cambridge University Press, Cambridge, 1912.

[6] Hilbert, D., Cohn-Vosson, S., *Geometry and the imagination* Chelsea, New York; 1952.

[7] Jared, G.E.M., Varady, T., Synthesis of volume modelling and sculptured surfaces in BUILD, In *Proceedings of CAD84,* Ed. J. Wexler, pp. 481-485, Butterworths, Guildford, 1984.

[8] Martin, R.R., Principal patches for computational geometry, Ph.D. Thesis, Cambridge University Engineering Department, Cambridge, 1982.

[9] Nutbourne, A.W., The solution of a frame matching equation, these proceedings.

[10] De Pont, J.J., Essays on the cyclide patch, Ph.D. Thesis, Cambridge University Engineering Department, Cambridge, 1984.

[11] Sederberg, T., Anderson, D., Goldman, R., Implicit representation of parametric curves and surfaces, Computer Vision, Graphics and Image Processing, 28, pp. 72-84, 1984.

[12] Varady, T., Personal Communication, 1984.

RECURSIVE DIVISION

M.A. Sabin
(Fegs Ltd., Cambridge)

Recursive division has two important roles in modern numerical geometry. One is an interrogation technique, where it gives a robust method for calculating sections and intersections. The other is as a definition procedure, where it provides a method of definition of surfaces with much more freedom in the logical positioning of the control points than previous methods. This paper covers both aspects. Section 1 gives the history of each aspect, Section 2 deals with interrogation techniques, and Section 3 covers the new class of surfaces described by these methods. Section 4 outlines the properties, both mathematical and operational, of the surfaces so far described in the public domain, and Section 5 concludes by indicating how many new surfaces are still to be discovered.

1. HISTORY

Warnock hidden surface algorithm

The first recognizable parent of the recursive division interrogation approach was the Warnock hidden surface algorithm (Warnock [18]). This was one of the first image space hidden surface algorithms, and certainly one of the first to perform significantly better than the 'compare everything with everything' methods previously in favour. The method has the very simple algorithm:

```
draw   scene   (picture) ::=
begin  if      picture is simple enough
       then    draw it
       else    if   picture is only one pixel
               then colour that pixel
               else divide picture into four quadrants
                    for each quadrant
                    do     draw scene (quadrant)
                    od
               fi
       fi
end
```

The division is of the picture into quadrants, and the recursion is in the calling of the procedure by itself.

This method turned out to be surprisingly fast for many scenes: it was discovered that large areas of many pictures are highly coherent, and could be painted in a single colour at a relatively early stage in the division.

Chaikin's curve

The first construction of the recursive division class was described at the 1974 Utah conference on Computer Aided Geometric Design (Barnhill and Riesenfeld [27]) by Chaikin (Chaikin [4]). His construction followed the method many carpenters have used to produce curved shapes, but made it a precise algorithm.

Given a polygon, construct another polygon by cutting off the corners of the first. This cutting off is to be done by joining the point three-quarters of the way along one edge to the point one quarter of the way along the next.

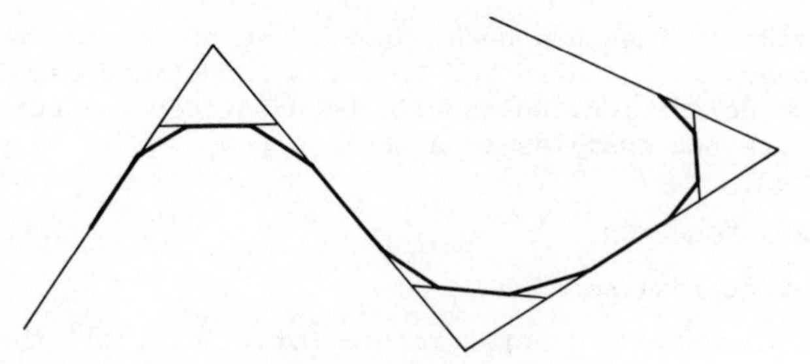

Fig. 1 Chaikin's construction

This produces another polygon, which can have the same procedure applied to it. Each application of the procedure gives a polygon with more sides than its predecessor, and which turns through a smaller angle at each corner.

When the lengths of all the sides are less than the resolution of the plotter there is little point in continuing, so stop.

One of the major new ideas at that conference was the B-spline as a curve design tool. It was soon observed that Chaikin's construction had a limit curve, and that this limit curve was the quadratic B-spline, a continuous curve consisting of a sequence of parabolae, each tangent to the next (and to the chord) at the midpoint of a chord of the original defining polygon (Forrest [7], Riesenfeld [14]).

Exercise 1 Prove that the limit curve of the Chaikin construction is the quadratic B-spline.

Exercise 2 Find the limit curve of a modified construction where corners are cut off by joining the point two-thirds of the way along one edge to the point one third along the next. (Hint: This is difficult and subtle. Do not waste too much time on it.)

Subsequent interrogations

The Warnock idea has spawned numerous methods in computer graphics (Sutherland, Sproull and Schumaker [12]). Some of these are image space methods, which divide a notional screen. Others are object space methods, which divide either the finite region of space the object of interest is located in or the logical extent of the object itself. The current state of development of this idea is in the various quad-tree methods for display (Oliver and Wiseman [10]), and in the oct-tree methods for spatial indexing in geometric modelling (Woodwark and Quinlan [19]).

Subsequent Constructions

Chaikin's idea has been generalised to construct B-splines of any order. It has also led to two well defined methods for surface definition, which will be described in Section 3 below. These are but examples of a whole class, so far very little explored.

2. INTERROGATION

Curve-line intersection

Here is a very general method for finding all the intersection points of a curve with a given straight line:

```
intpoints (curve, line) ::=
begin if    curve is simple enough
      then  calculate any intersection points
      else  if     curve satisfies test for no intersections
            then   no intersection points
            else   divide curve into two halves
                   union (intpoints (first half, line),
                          intpoints (second half, line))
            fi
      fi
end
```

Applications of this method to different curves will vary in four places only:-

(i) the test for simplicity

(ii) the direct calculation in the simple case

(iii) the test for definite absence of intersections

(iv) the halving procedure

The first of these is normally a test to see whether the curve is adequately approximated by a straight line, in which case the second is just a straight line/straight line intersection .

The third is most easily available if the curve has a Bezier or B-spline definition, when the overlap of the convex hull of the control points with the line can be tested for without actually calculating the convex hull itself. One of the useful properties of these forms is that the curve is guaranteed to lie within the convex hull of the control points. However, the basic algorithm is not restricted to such curves. As long as some way of testing bounds exists, or better, some way of constructing a set of points within whose convex hull the curve lies, the same outer algorithm will work.

The fourth is the place where the actual equation of the curve is important, since this is where the algorithm for creating the control points of the halves from those of the original depends on the equation.

Exercise 3 How can the test for overlapping of the convex hull avoid forming the convex hull? (Hint: the intersection is with a straight line.)

The basic algorithm is guaranteed to give all the points of intersection, and if the union process is implemented sensibly will give them in the right order. Similar algorithms cover the problems of:-

* curve-curve intersections (in two dimensions)

* surface-plane intersections (in three dimensions)

and

* surface-surface intersections.

The last of these has the algorithm:

```
intcurve (surface 1, surface 2) ::=

begin  if    both surfaces are simple enough

       then  calculate any pieces of intersection curve

       else  if    surfaces satisfy test for no intersections

             then  no intersection to calculate

             else  choose which surface to divide

                   choose which way to divide it

                   union (intcurve (first half, other surface),

                          intcurve (second half, other surface))

             fi

       fi

end
```

This algorithm has additional complications which are:

(i) that the test for no intersections cannot avoid the convex hull construction quite so simply. The usual solution is to use a bounding box instead of the convex hull. If the bounding boxes of the two sets of control points do not overlap the surfaces cannot intersect. This is a looser test than the convex hull one, but there is evidence that its simplicity makes it a good buy.

(ii) that the choice of which surface to divide can be critical in determining the performance of the implementation. Simple rules are to avoid dividing a simple surface, and otherwise to divide the larger of the two, for example that with the larger bounding box.

(iii) that choice of the direction of division is also important. The simple rule of dividing the longer dimension is probably as good as any.

(iv) that the union now takes a little bit of thought if the pieces of curve constructed by the lower level calls are to be put together in curve order.

A sequence of Ph.D students at University of East Anglia have addressed all these problems and have found an adequate supply of solutions (Palmer [11], Geisow [8], Peng [12], Nasri [9]).

3. CONSTRUCTIONS

If a surface is to be interrogated by the algorithms of Section 2, we may view the elemental algorithms, particularly the halving process, which derives the control points of the half surfaces from those of the original, as being the definition of the surface.

The Catmull Clark cubic

Given a polygon we may construct a B-spline curve by the following cubic analogue of Chaikin's algorithm

```
begin

    for each edge in the control polygon

    do   construct the midpoint

    od

    for each vertex in the original control polygon

    do   construct the point at the mean of (that vertex) and
              (the mean of the two midpoints just constructed
                   adjacent to it).

    od

    create a new polygon by taking the newly created points in
              the obvious order, alternating with the edge
              midpoints.

end
```

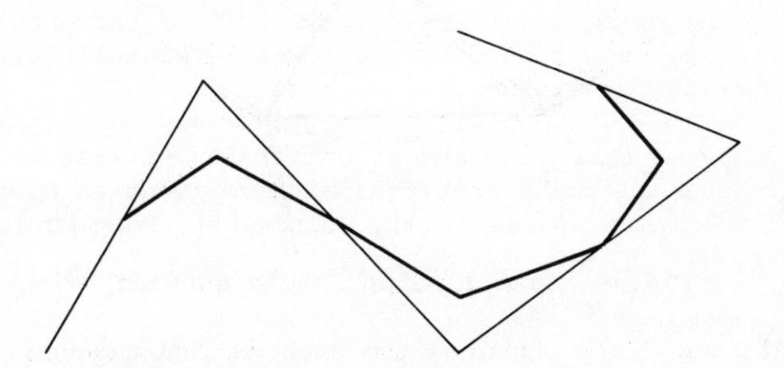

Fig. 2 Cubic analogue of Chaikin

The tensor product of two of these constructions gives a procedure which constructs new vertices at the centroid of each face in the network of control points, then new points corresponding to the midpoints of the edges of the net, and then finally new points corresponding to the original control points. The new net joins the newly constructed points in the obvious way, with four new faces replacing each original face of the net.

Catmull and Clark generalise this surface algorithm as follows. A new point can be constructed at the centroid of a face of any number of sides, the construction of a new edge midpoint applies as before, and a new vertex can be calculated by some weighted mean or another of the old vertex with the centroid of the adjacent new edge points.

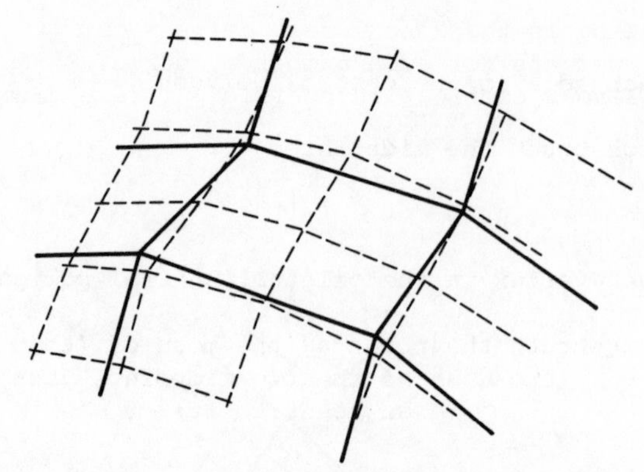

Fig. 3 Tensor product of cubic Chaikin

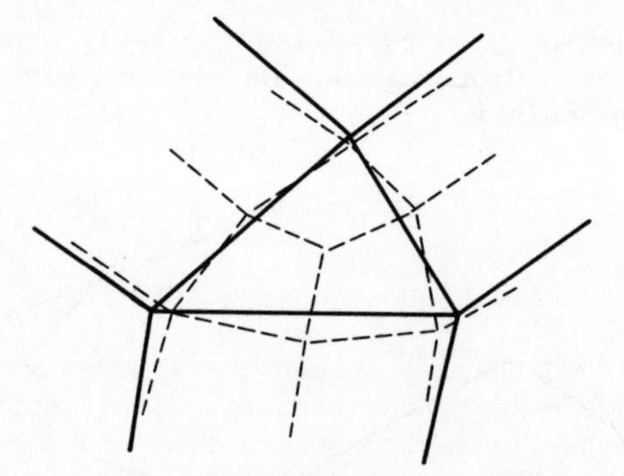

Fig. 4 Catmull-Clark cubic surface

Catmull and Clark explored two such weighted means, finding by experiment that the first one did not give perfect pictures (Catmull and Clark [3]).

Exercise 4 How might you choose a good ratio for the vertex points?

The Doo-Sabin quadratic

We have shown that the recursive division of a cubic B-spline curve can give a bicubic B-spline by taking the tensor product, and that this process can then be generalised to irregular bivariate meshes, Chaikin's construction itself can generate a biquadratic B-spline, but it is not immediately clear how to generalise the following resulting construction.

Create a new face inside each old face, using the linear combinations (1:3:9:3)/16 of the old vertices to construct the new ones. Join these faces across the old edges by new faces, and fill in across the old vertices by new faces too.

The solution is that one merely defines new linear combination weights for each number of edges, n say, round a face. The linear combination coefficients are given (Doo and Sabin [5]) by

$$w_{ii} = (n+5)/4n$$

$$w_{ij} = (3 + 2\cos(2\pi(i-j)/n))/4n \quad i \neq j$$

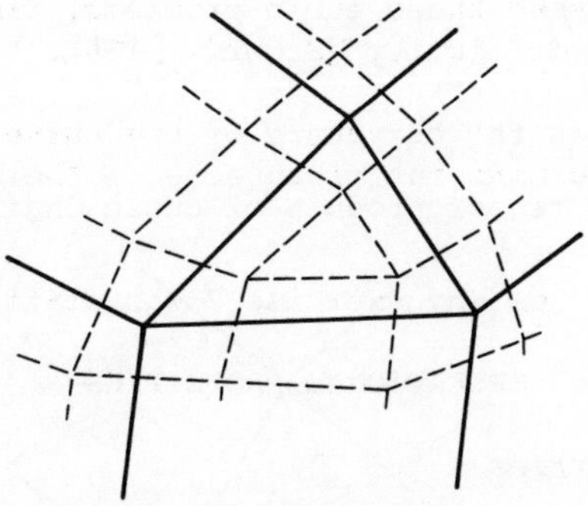

Fig. 5 Doo-Sabin quadratic

We also observed that, provided the 1:3:9:3 ratios for the four-sided face are kept, the resulting limit surface had biquadratic patches in regions where four-sided faces met at a vertex, and thus the construction was a true generalisation of the quadratic B-spline.

Of more interest was the analysis of the behaviour near the singular points which occur at the centres of non-four-sided faces. The configuration of control points around such a face after an iteration can be put into one-one correspondence with the configuration before. The new points are all linear combinations of the old, and so we may regard the construction as applying a linear operator to the configuration. What happens after many iterations depends on the result of a high power of the linear operator, which is analysable in terms of the eigenproperties of the operator.

The largest eigenvalue is unity, which corresponds to all control points being equal. This indicates that the configuration as a whole stays in much the same place and does not drift off to infinity. The second distinct eigenvalue, λ say, gives the rate at which the configuration shrinks in its eventual tangent plane. The third distinct value, μ say, controls the rate at which the out of plane components shrink towards the tangent plane. Any estimator of curvature is multiplied at each step by μ/λ^2 which must therefore be equal to unity for good curvature behaviour.

It was found that the quadratic behaves well at singular points. Any path toward a singular point approaches it with essentially constant curvature in the limit. Further, the misbehaviour Catmull and Clark had observed in their first cubic was explained by their corresponding curvature increasing without limit. Their second cubic had its curvature collapsing toward zero, but relatively slowly. Ball and Storry [1] have subsequently addressed these cubic problems, finding weights which minimise the difficulty (Storry [16]).

Exercise 5 How does the curvature of the curve of exercise 2 above behave as the midpoint of an edge of the original polygon is approached?

Exercise 6 Do you see why exercise 2 was difficult?

4. MATHEMATICAL AND OPERATIONAL PROPERTIES

Mathematical properties

All the recursive division definitions result in a structure, after many divisions, which has large regions of some regular network of control points, with just a few singular points at which a non-regular number of edges meet at a vertex or round a face.

It is possible, but not always true, that each regular piece of control network (of a size determined by the region of support of the effect of moving one of the original control points) will have an analytic limit surface.

If this does happen, the surface round each singular point will consist of successive rings of analytic pieces at successively smaller scales. There will be an infinite total number of pieces, but almost all of them contribute almost no area to the surface.

Operational properties

A surface is defined by a network of control points not necessarily lying in the surface, B-spline fashion. All of the good ergonomics of the B-spline are retained by this extension.

The network is almost unconstrained in its connectivity, the only rules being that each face must have at least three vertices, and each vertex at least three incident edges. This makes the new method much more suitable for complicated shapes with many feature lines crossing each other in awkward ways.

If the natural smoothness is exploited in the use of a very coarse network to control regions of the surface without short-wavelength features, the quadratic tends to bulge into the corners. The shape is not therefore quite ideal. To fit a sphere closely needs a control polygon two levels of division finer than the cube.

5. OTHER CONSTRUCTIONS

The two constructions described above are but samples of the possibilities. A few more, not explored yet in detail, may give some idea of the riches yet to be discovered, and the methods which might be used in such discovery.

Consider the construction:-

(i) for each original face, construct a new face by joining the midpoints of the edges of the old.

(ii) for each original vertex create a new face by joining the midpoints of the edges meeting at that vertex.

What is the limit surface, and how does it behave near its singular points?

This is reminiscent of exercise 2, but there is a systematic approach which can tease out many properties, given only the construction.

(i) After the first iteration all vertices are 4-valent: all new vertex-faces have the same number of sides as the original vertex had spokes, and each new face-face has as many edges as its original. In the limit the configuration has large areas of regular rectangular mesh, meeting at isolated singular points.

(ii) The regular mesh rotates through 45 degrees at each iteration, shrinking by a factor of sqrt(2).

(iii) The basis functions over the regular grid have a region of support octagonal in shape. (This is identified by starting with a single non-zero z-value and seeing how the non-zeros propagate out as the iterations pass. Once the basic pattern is visible a series can be set up which is easily summed to give the limit of the support region.

(iv) By superimposing these octagons around all the mesh points in a sufficiently large region, we see that if the limit surface has an algebraic expression it must be piecewise with a partitioning into triangular pieces arranged in a St. Andrews cross arrangement. (This partitioning is similar to that used by (Sibson and Thompson [15]).

(v) If all the mesh points in or on the boundary of a support octagon lie in a plane, those lying in a region larger than one such octagon (at the next level of division) will lie in the the same plane after division. After many divisions a piece of mesh not tending to zero area will lie in that plane. Planarity is thus stable, and so the precision set of the surface includes the plane.

(vi) Similarly, if all mesh points in a support octagon lie on the quadratic z=x squared, there is a non-zero quadratic area after many divisions. By symmetry and linearity this will hold for any quadratic function. The precision set thus includes quadratic functions.

With these clues we can confirm that the limit surface does in fact consist of quadratic functions over the triangles indentified in (iv) above. These pieces meet with continuity of slope, but obviously not of curvature.

In fact this surface may be described as a multivariate B-spline, (a box-spline), the basis function being the shadow of the projection of a 4 dimensional cube projected onto the two abscissa dimensions. (Prautzsch [13]). (Also see the paper by Boehm in these proceedings).

<u>Exercise 7</u> How does the surface behave around a singular point? (Hint: it is easier to deal with the square of the division operator, as the net rotation can be treated as zero and the asymmetry of a single operator avoided).

<u>Exercise 8</u> What is the limit surface of the construction:-

(i) At the centroid of each old face create a new face-vertex.

(ii) Make a new vertex-vertex halfway between each old vertex and the mean of the new face-vertices surrounding it.

(iii) Make new (triangular) faces by joining each new vertex-vertex to each adjacent pair of new face-vertices around it.

<u>Exercise 9</u> What is the limit surface of the construction:-

(i) and (ii) as exercise 8

(iii) make new (four-sided) faces by joining the vertex-vertices corresponding to the ends of each original edge to the face-vertices on either side of that edge.

Exercise 10 What is the limit surface of the construction:-

(i) Create a new face inside each old face by constructing vertices as linear combinations of the old.

(ii) Create a new (six-sided) face across each old edge by joining the new face-vertices and the new vertices halfway between the old vertex and the centroid of the new vertices which surround it (Farin [6]).

REFERENCES

[1] Ball, A.A. and Storry, D.J.T., Recursively Generated B-spline surfaces, *Proc. CAD84*, pp. 112-119, Butterworths, 1984.

[2] Barnhill, R.E. and Riesenfeld, R.F. (eds), *Computer Aided Geometric Design*, Academic Press, 1974.

[3] Catmull, E. and Clark, J., Recursively generated B-spline surfaces on arbitrary topological meshes, CAD vol. 10, pp. 350-355, 1978.

[4] Chaikin, G., An Algorithm for High Speed Curve Generation, Computer Graphics and Image Processing, vol. 3, 346-349, 1974.

[5] Doo, D. and Sabin, M., Behaviour of recursive division surfaces near extraordinary points, CAD vol. 10, pp. 356-362, 1978.

[6] Farin, G., Designing C^1 surfaces consisting of triangular cubic patches, CAD vol. 14, pp. 253-256, 1982.

[7] Forrest, A.R., Notes on Chaikin's Algorithm, Computational Geometry Project Memo CGP74/1, University of East Anglia, 1974.

[8] Geisow, A., Surface interrogations, Ph.D. Thesis, University of East Anglia, 1983.

[9] Nasri, A., Polyhedral Subdivision Methods for Free-form surfaces, Ph.D. Thesis, University of East Anglia, 1984.

[10] Oliver, M.A. and Wiseman, N.E., Operations on quadtree leaves and related image areas, Computer Journal vol. 26, pp. 375-380, 1983.

[11] Palmer, T.R., Sculptured Surfaces in Volume Modelling Systems, Ph.D. Thesis, University of East Anglia, 1982.

[12] Peng, Q.S., Volume Modelling for Sculptured Surfaces, Ph.D. Thesis, University of East Anglia, 1983.

[13] Prautzsch, H., Unterteilungsalgorithmen für multivariate splines, Doctoral Thesis, Universitat Braunschweig, 1984.

[14] Riesenfeld, R.F., On Chaikin's Algorithm, Computer Graphics and Image Processing, vol. 4, pp. 304-310, 1975.

[15] Sibson, R. and Thompson, G.D., A Seamed Quadratic Element for Contouring, Comp.J. vol. 24, pp. 378-382, 1981.

[16] Storry, D.J.T., B-Spline surfaces over an irregular topology by recursive subdivision, Ph.D. Thesis, Loughborough University, 1984 (to be submitted).

[17] Sutherland, I.E., Sproull R.F. and Schumaker, R.A., A Characterisation of Ten Hidden-Surface Algorithms, ACM Computing Surveys 6, 1974.

[18] Warnock, J.E., A Hidden Surface Algorithm for Computer Generated Halftone Pictures, University of Utah Computer Science Department, TR4-15, 1969. NTIS AD-753 671.

[19] Woodwark, J.R. and Quinlan, K.M., The derivation of graphics from volume models by recursive division of the object space, *Proc. CG80*, Online Publications 1980.